Local Government and Sustainable Development in the UK

Drawing on and advancing theoretical frameworks that emphasise risk and reflexive modernity, this book presents data, discussions, and recommendations from the UK to explore sustainable development and its associated risks at the local, community, and individual levels. It argues that local government provides an underused link between communities and international governance that can directly address some of the most pressing problems that face society today.

The focus of the research and analysis presented in this book is the local government tier of parish councils particularly drawing on those from the county of Cornwall, UK. There are over 10,000 town and parish councils in England and Wales that sit at the heart of many local communities. They are the closest level of government to their communities and play a pivotal role in shaping these communities. As such, this book puts forward the need for a reconceptualisation of parish councils that not only sit within local governance structures, but also are impacted by and impact upon global governance discourses. Through the medium of parish councils, the author explores how local governments globally can contribute to the impact of the sustainable development goals on communities around the world.

This book will be of great interest to students and scholars of sustainable development, governance, and political communication.

Gregory Borne is a Course Leader and Academic at Falmouth University. Borne has over 25 years of experience in researching, teaching, and implementing multiple dimensions of sustainable development. He has worked with international organisations such as the United Nations as well as national and local government, business, and the third sector. Borne is a chartered environmentalist, a member of the Royal Institution of Chartered Surveyors, and a member of the Institution of Environmental Sciences.

Routledge Studies in Sustainable Development

This series uniquely brings together original and cutting-edge research on sustainable development. The books in this series tackle difficult and important issues in sustainable development including: values and ethics; sustainability in higher education; climate compatible development; resilience; capitalism and de-growth; sustainable urban development; gender and participation; and well-being.

Drawing on a wide range of disciplines, the series promotes interdisciplinary research for an international readership. The series was recommended in the *Guardian*'s suggested reads on development and the environment.

Inside the Sustainable Development Goals
From Fighting Poverty to Saving the World
Adele Orosz

Circular Economy Solutions for Sustainable Development
Katarzyna Nowicka

How to Achieve Sustainable Development
Local Actions for Global Change
Samuel T. Ledermann

Innovation Strategies and the Sustainable Development Goals
A Blueprint for Sustainable Transformation
Amit Singh Khokhar, Punita Duhan and Anshu Lochab

Artificial Intelligence for Sustainable Development
How, Why, and What Drives the Future
Medani P. Bhandari

For more information about this series, please visit: www.routledge.com/Routledge-Studies-in-Sustainable-Development/book-series/RSSD

Local Government and Sustainable Development in the UK

Engaging the Local to Achieve the Global

Gregory Borne

Routledge
Taylor & Francis Group
LONDON AND NEW YORK

earthscan
from Routledge

First published 2026
by Routledge
4 Park Square, Milton Park, Abingdon, Oxon OX14 4RN

and by Routledge
605 Third Avenue, New York, NY 10158

Routledge is an imprint of the Taylor & Francis Group, an informa business

British Library Cataloguing-in-Publication Data
A catalogue record for this book is available from the British Library

ISBN: 978-0-367-27663-8 (hbk)
ISBN: 978-1-041-25069-2 (pbk)
ISBN: 978-0-429-29719-9 (ebk)

DOI: 10.4324/9780429297199

Typeset in Times New Roman
by codeMantra

This book is dedicated to my father, John.

Contents

	Prologue	*ix*
	Acknowledgements	*xi*
1	Introduction	1
2	Sustainable Development and a Risk Society	8
3	Local Governance in a Global Context	27
4	Parish Context	48
5	Sustainable Development and Risk	72
6	Internalising Sustainable Development	92
7	The Salience of Risk	109
8	Sustainable Communities	129
9	Identifying the Local to Achieve the Global	147
10	Sustainable Developmental and Reflexive Modernity Connecting the Global to the Local	164
	Appendix1: Questionnaire	*177*
	Index	*187*

Prologue

I was born on the South Coast of Devon, in an Area of Outstanding Natural Beauty, what is now called a National Landscape, and my family heritage is scattered throughout the counties of Devon and Cornwall. I have also lived my whole life within sight, sound, and smell of the ocean. The Cornish word for the ever-present sound of the ocean is *mordros,* and as this book unfolds we will see the significance of the ocean to the research context.

My family were farmers in both counties, in Mousehole and Penzance in Cornwall, as well as Start Point and Ringmore, in the South Hams in Devon. The house I grew up in was once a barn on my maternal great-grandfather's farm, passed to my grandmother, Narissa-Anne, and subsequently to my parents, John and Jenny. With historical records tracing the original structure of the property to as early as 1000 AD, I have always felt a strong sense of connection to the house and the surrounding village. The building was renovated over the years by my paternal grandfather, Sydney, who lived in our family home and was a stone mason and builder by trade. I still feel a strong sense of loss after the decision was made to sell the property following my father's death. During my childhood and adolescence, the village I grew up in had five working farms, and at one time or another I played and worked on all of them. Today none of those farms remain, with land being sold and property converted.

There is little doubt that the natural beauty of the environment that I was privileged to grow up in has had a profound effect on my life, both personally and professionally. I have gravitated towards issues relating to sustainable development, and over the years I have meandered through different aspects of this concept, with no particular plan or direction, simply pursuing issues of interest and inspiration.

This potted personal history is of direct relevance to the work presented in this book, a book that explores rural voices and their connection to the concept of sustainable development. Many of those voices resonated strongly with my own experiences, both reinforcing and challenging my understanding of the world. I am therefore not separate from these voices but inextricably linked to the narratives and issues discussed throughout this book.

I will also add that whilst extensive research was conducted in both Devon and Cornwall, this book has focused very directly on the Cornish component of the research. And whilst I am proud of my Devonian heritage, circumstances and academic coherence have dictated that Cornwall drives the direction of this book. It is my hope that in some small way the observations presented here help to contribute in a positive way to communities across Devon and Cornwall.

And in no small measure this book also represents a reconnection with my own Cornish heritage. This book has been a long time in the making, with many false starts and circumstances dictating on many occasions that the time was not right to take the work forward. However, having spent most of my life in Devon, at the time of writing I am now an academic at Falmouth University, Cornwall, which has perhaps provided the right set of circumstances to continue making the professional and personal journey that has resulted in the completion of this book.

My paternal great-great-grandfather, Robert Borne, was the head constable of Falmouth, his son John was born in Falmouth, and, later a submariner in the First World War, eventually moved his family to Modbury in the South Hams, Devon, the area where I grew up. Finalising this book feels like a journey home in many ways. Perhaps most importantly of all, reconnecting and understanding his Cornish heritage mattered deeply to my father, and I am proud to in some small way to take this legacy forward here.

Acknowledgements

My family, Janine, Sydney, and Sophia, for their continued and unwavering support, through good times and bad.

1 Introduction

This introductory chapter will provide the rationale and overview for this book. As the title of this book demonstrates, this book is about sustainable development. But it is about the fluidity, ambiguity, and contradictory nature of the concept. It is as much about what sustainable development is not as it is an explanation of what the concept means. There are multiple textbooks and sources that can be consulted for an overview of the concept as well as methods and frameworks for applying the principles of sustainable development in policy and practice (Becker 2023; Blewitt 2017; Roorda 2025; Rogers et al. 2012; Lafferty 2006; Dalby et al. 2019; Orosz 2025; Grin et al. 2010). This book is about context, it is about connecting the global to the local, it is about narratives and discourse, and it is about the construction of the concept of sustainable development and multiple associated issues. It is also about exploring a theoretical framework that is capable of informing multiple dimensions of sustainable development whilst simultaneously utilising discourses of sustainable development to text the very same theoretical framework.

This book builds on empirical observations in the context of town and parish councils in the county of Cornwall, UK. This book explores individual, community, and political narratives that enable an in-depth exploration of past and current policy-making, as well as identifying barriers for constructively embedding sustainable development both locally and globally. And whilst this book's context is firmly embedded in the local domain, it is also an examination of how local dynamics can inform global strategy and policy, and vice versa.

This book elaborates on and takes forward the relationship between theory, empirical work, and practice that is part of a continuum that began with my doctoral thesis in 2002. This continuum has meandered through nearly more than two and a half decades, exploring how the concept of sustainable development has moved from the political, economic, environmental, and social sidelines to now being a central concept on the socio-political, economic, and environmental stage. This work has been underpinned by ongoing theoretical observations that are grounded in an exploration of the relationship between sustainable development and the German sociologists Ulrich Beck's Risk Society Thesis (Beck 1992, 1999; Borne 2006, 2010, 2018; Borne and Ponting 2017). I have argued elsewhere that there is a symbiotic relationship between the rise of sustainable development on the socio-political stage and the emergence of what Beck describes as a reflexive modernity. That is, a modernity that

DOI: 10.4324/9780429297199-1

now challenges some of the central tenets of the modernist agenda. This includes, though is not limited to, progress through science and technology, certainly of knowledge, control over nature, industrial growth and economic expansion, institutions, and the autonomy of the nation state. Ultimately a reflexive modernity is a modernity that is turning in on itself. Chapter 2 will elaborate on these issues, making connections between theory and practice, global and local, and the relationship between sustainable development and a risk society.

The connection between the global and the local, the transmission of sustainable development messaging, as well as institutional development, policy, and legislation from an international and national level, and the translation of these broader structural frameworks at the local level remains a consistent theme throughout this book. But instead of exploring the transmission of global narratives to the local level, we start at the local, community, and individual level, and explore the relationship this has to global processes.

The data that informs the central narrative of this book was collected during an extended period between 2007 and 2010. This timescale provides a unique opportunity to unpack the complex narratives of sustainable development and connect the political and policy landscapes from that time to the present day. The issues identified within the research have accelerated in importance and prominence over the past 15 years, and it is instructive at this point to make initial connections between then and now.

The return of the Labour government in July 2024 under Prime Minister Keir Starmer provides a political and ideological bridge between the time of the data collection and today. Whilst this is not the same Labour government that existed before 2010, indeed it would be profoundly misleading to directly map the current Labour government to that which governed prior to 2010 (Wills 2016). The return of this Labour government and the connection to the previous Labour government provide an opportunity to explore the changing political landscape. This temporal analysis also enables us to look forward to exploring policy trajectories and understanding cycles, themes, successes, and failures. Furthermore, making these connections provides the ability to explore whether past experiences have helped shape present-day expectations and localised governance strategies.

Historically, Labour governments have promoted community-led sustainability, public sector investment, and climate-focused policy interventions (Cowell et al. 2011). This provides an opportunity to reengage with policies and ideologies of devolution that present inconsistent and often contradictory messaging about the ways that principles of subsidiarity are understood. As Wills points out we need to be cautious about absolutist assumptions that present polemics of centralist and devolutionist approaches.

> Whereas the New Labour governments (1997–2010) sought to use the state to engage with the people, the previous and current Conservative-led governments have been more liberal and less prescriptive, passing permissive legislation that creates the space for citizens to act on their own terms and in their own way.
>
> (Wills 2016:4)

Wills explains that instead of freeing local government to take their own reigns and chart their own priorities responding to local imperatives, New Labour '… implemented tighter control over its work, whilst at the same time encouraging contracting with the private sector and charitable organisations and setting up wholly new structures for the delivery of some strategic ideas' (Wills 2016:66). And this expansion of the governance process reflected broader trends globally led by countries within Europe as well as other organisations such as the World Bank. Wills continues to evaluate the history of local government and its role in the public realm. Perhaps most importantly for the narratives, analysis, and conclusions drawn in this book, Wills tells us that '…the paradox of English local government, greater democratisation was associated with ever- greater centralisation' (2016:70). This book directly addresses this observation through embedded local narratives, evolving policy landscape, and the heightened focus on sustainable development. In Chapter 10 we will revisit this evolution in light of the empirical work discussed and the way this relates to the implementation of the sustainable development goals (SDGs) (UN 2025)

The discussions in this book provide timely and critical insights into the past challenges, policy responses, and implementation barriers that are still clearly visible today. Many of the structural and policy issues identified in the research remain locked in, and this book provides a conduit through which we can move back and forth between policy landscapes to explore realistic solutions informed by the lessons of the past. Key barriers, such as funding constraints, multilevel governance conflicts, and local autonomy limitations, continue to impact local government effectiveness and ultimately sustainable development implementation (Pike et al. 2016).

As a result we are able to track persistent barriers and contradictions that contextualise sustainable development intervention. We are also able to explore whether policy interventions since 2010 have meaningfully addressed these challenges, as well as shedding an urgent light on scenarios where there have been significant policy stagnation, not to mention failures. This is a timely analysis with the Public Administration and Constitutional Affairs Committee Report (PACA 2024) identifying an urgent need to reform local government in England. The key rationale for this finding is to strengthen and restore public trust, as governance structures have been identified as being complex, confusing, opaque, overcentralised, and underfunded. Ultimately the committee found that

> People in England have a low sense of political efficacy, meaning that they do not feel political and social change is possible, and in particular they feel that their participation in the political process is unlikely to bring about change.
>
> (PACA 2024:4)

In part, as a result of this finding in December 2024 the government published The English Devolution White Paper, 'Power and Partnership: Foundations for Growth'. The paper indicates that

> Policies have been done to communities, rather than for and with them, with one-size-fits all approaches and complicated funding processes that reflect

> the silos of Whitehall rather than the needs of communities. This has left people feeling as if their lives, and the places they call home, have slipped out of their control.
>
> (MHCLM 2024)

In this context Devolution '…refers to the handing over of certain powers usually exercised by central government to lower levels of government' (POST 2025). Through an in-depth exploration of key issues at the local government level we will be able to not only address these observations but also expose some of the underlying assumptions upon which it is built.

Foregrounded by the above there remains a dearth of research that is capable of addressing perceptions, narratives, and discourses of sustainable development. The ways in which individuals and communities interpret sustainable development and associated risks such as global warming are critical to understanding policy effectiveness and governance legitimacy. Whilst sustainable development mechanisms have evolved, especially in relation to the SDGs, core debates persist. This includes the tension between local and global priorities, perceived responsibilities for problem framing and solutions, not to mention public trust in frameworks and mechanisms for achieving a sustainable future. The research presented in this book provides a unique snapshot of local communities' perceptions of key issues relating to sustainable development for the UK and helps inform the world's trajectory towards sustainable development.

Sustainable development is not static; it evolves in response to political shifts, global policy commitments, and technological advancements. However, many of the structural barriers and governance contradictions identified in the research presented in this book remain fundamentally unresolved. Themes and issues identified in this book enable us to track governance trends over time, providing a lens through which to evaluate the effectiveness of past policy interventions. These observations also enable us to critically examine whether sustainable development commitments have resulted in meaningful local action. And more generally we are able to explore the long-term impact of political, economic, and technological shifts on local governance. Ultimately, the research not only highlights what has changed but, more importantly, what has remained the same, offering valuable lessons for policy-makers, scholars, and practitioners working towards more effective and equitable sustainability governance today.

Book Structure

This book is structured in the following way. In Chapter 2, I will discuss sustainable development and the relationship the concept has to a risk society and particularly a reflexive modernity. In doing so the theoretical underpinnings of this book are established, and this will be a consistent theme throughout this book. This chapter will also explore the evolution of the concept of sustainable development and ultimately explore the emergence of the United Nations Sustainable Development Goals (UNSDGs) that were established in 2015.

Chapter 3 focuses on the policy and legislative landscape that town and parish councils inhabit. Taking forward the theoretical propositions introduced in Chapter 2, utilising a sustainable development lens as well as drawing on a multilevel governance framework to structure the discussion, this chapter explores the interconnections between multiple relevant themes. This chapter will also chart the changing socio-political and economic landscape from the initial research period through to the present day.

Chapter 4 presents a methodological overview and proceeds to discuss initial data, predominantly focused on the survey results. Identifying some key characteristics of Cornwall, this chapter sets the scene for the context of both the survey and the interviews and establishes initial perceptions and understandings of key areas addressed in this book. This includes the initial demographics of the research participants, as well as exploring the priorities and perceptions of town and parish councillors at that time. Understating notions of urban and rural, partnerships and connections to higher tiers of local government are highlighted. Broad perspectives on community are also explored as well as identifying interpretations of sustainable communities. This chapter also identifies barriers to town and parish councils' ability to create sustainable communities, especially through key factors that cause problems for effective service provision.

Chapter 5 builds on the broad insights established in Chapter 4, presenting further survey data and infusing this where appropriate with more in-depth observations from the interview data. This chapter continues to build our understanding of community and deepens the engagement with sustainable development. This chapter also explores awareness of local governance mechanisms for achieving sustainable development and sustainable communities. This chapter also establishes the relationship between sustainable development and risk by identifying a number of risks that are identified as part of the broader narratives relating to sustainable development: issues such as climate change and global warming, and ozone layer depletion. This chapter presents findings of perceptions of sustainable development, starting with the three pillars of sustainable development, environment, economy, and society and progressively building on this to explore consistencies and contradictions in sustainable development discourse. Whilst this chapter firmly locates sustainable development within a Cornish context, it also explores town and parish councillors' global narratives and reflections, making direct connections between the global and the local scales of analysis. Together Chapters 4 and 5 present a robust picture of town and councils' understanding of and engagement with multiple dimensions of sustainable development.

Chapters 6–8 present discourse and analysis drawn from the interviews with town and parish councillors. And whilst the dominant emphasis of each chapter is distinct, sustainable development, risk, and community respectively, in reality, what we witness is the fluid and interconnected nature of the discourses that relate to sustainable development. In Chapter 6 the interviews present a rich and layered understanding of multiple issues relating to sustainable development. There is an overarching focus on sustainable development discourse within this chapter, but as we shall see this chapter demonstrates the intertwined nature of associated narratives. Indeed, this chapter explores reasons why unpacking perceptions of sustainable development presents challenges for academic investigation, but also benefits from the theoretical grounding of a risk society.

Chapter 7 focuses on discourses of risk which inevitably draw together the discourse of community and sustainable development, making essential connections to the underlying theoretical framework. This chapter highlights how councillors seem to ground their understanding of sustainable development within the context of stewardship and heritage, bypassing perceived abstract policy language. Overall, this chapter frames sustainable development through local cultural logics, contributing to the argument that sustainable development discourses are contextually negotiated and not universally applied.

Chapter 8 focuses very directly on the discourse of community and sustainable community. This chapter explores connections to place, nature, heritage, and belonging, and indeed the loss of these things. This chapter also explores councillors' understanding of governance processes as well as key issues that impact the multifaceted concept of community. This chapter highlights councillors' understanding and response to agricultural challenges such as housing and demographic change as well as reflecting on how councillors understood their roles within town and parish councils. Altogether this chapter suggests how community identity acts as both a driver and a constraint to local engagement with sustainable development. Moreover, this chapter shows how a sense of place can form a context for engagement with reflexive modern processes.

Chapter 9 moves the focus of the discussion to the international level. This is achieved with a particular focus on the United Nations where narratives of sustainable development are discussed within the context of evolving sustainable development discourse and the implications this has for engagement with sustainable development and the SDGs. Structured through themes of divergence, convergence, dissolution, and evolution, this chapter demonstrates how sustainable development discourse has moved past an original focus on the environment to become a multifaceted space for addressing social, economic, and environmental issues. Also this chapter charts the shift in sustainable development discourse through key documentation and international events that have been responsible for disseminating the message of sustainable development from the global to the local level. Comprehensively, this chapter identifies sustainable development as both a catalyst for and evidence of a reflexive modernity and provides a way to connect the global and the local components of this book.

Finally, Chapter 10 draws together the global and local scales of analysis, applying and adapting components of reflexivity, expanding the relationship between sustainable development and a reflexive modernity. This chapter presents policy-relevant suggestions and advances theoretical debates as well as offering avenues for future research. Overall, by contextualising local experiences within the shifting dynamics of sustainable development governance as well as applying a reflexive modernity lens, the chapters that follow offer a unique perspective on how local communities engage with global risks.

Bibliography

Beck, U. (1992). *Risk society, towards a new modernity*. Sage.

Beck, U. (1999). *World risk society*. Polity Press.

Becker, P. (2023). *Sustainability science: Managing risk and resilience for sustainable development*. Elsevier.

Blewitt, J. (2017). *Understanding sustainable development*. Routledge.

Borne, G. J. (2006). Sustainable development: The reflexive governance of risk (Thesis, University of Plymouth).

Borne, G. (2010). *Framework for sustainable global development and effective governance of risk*. Edwin Mellen Press.

Borne, G., & Ponting, J. (Eds.). (2017). *Sustainable surfing* (p. 276). Routledge.

Borne, G. (2018). *Surfing and sustainability*. Routledge.

Cowell, R., Bristow, G., & Munday, M. (2011). Acceptance, acceptability and environmental justice: The role of community benefits in wind energy development. *Journal of Environmental Planning and Management*, *54*(4), 539–557.

Dalby, S., Horton, S., Mahon, R., & Thomaz, D. (2019). *Achieving the sustainable development goals. Global governance challenges*. Routledge Taylor & Francis.

Grin, J., Rotmans, J., & Schot, J. (2010). *Transitions to sustainable development: New directions in the study of long term transformative change*. Routledge.

Lafferty, W. M. (Ed.). (2006). *Governance for sustainable development: The challenge of adapting form to function*. Edward Elgar Publishing.

Ministry of Housing, Communities & Local Government. (2024). *English devolution white paper: Power and partnership*. Foundations for growth. GOV.UK. https://www.gov.uk/government/publications/english-devolution-white-paper-power-and-partnership-foundations-for-growth/english-devolution-white-paper

Orosz, A. (2025). *Inside the sustainable development goals: From fighting poverty to saving the world*. Routledge.

PACA. (2024). Governing England, third report of the session 2022–2023. https://committees.parliament.uk/publications/31418/documents/176171/default/

Pike, A., Rodríguez-Pose, A., & Tomaney, J. (2016). *Local and regional development*. Routledge.

POST. (2025). Community engagement with English devolution. https://post.parliament.uk/community-engagement-with-english-devolution-2/

Rogers, P. P., Jalal, K. F., & Boyd, J. A. (2012). *An introduction to sustainable development*. Routledge.

Roorda, N. (2025). *Fundamentals of sustainable development*. Routledge.

United Nations. (2025). Sustainable development goals. https://sdgs.un.org/goals

Wills, J. (2016). *Locating localism: Statecraft, citizenship and democracy*. Policy Press.

2 Sustainable Development and a Risk Society

This chapter will critically draw on the theoretical observations by Ulrich Beck as presented in his World Risk Society Thesis and explore its relationship with sustainable development. It will be argued that a risk society provides sophisticated and constructive theoretical base for exploring the complex issues that surround sustainable development. Initially the concept of sustainable development will be introduced with emphasis on the ambiguous and conflicted nature of the term. This will be followed by introducing the central tenets of Ulrich Beck's Risk Society. This initial section will conclude by drawing four primary synergies between sustainable development and globalised world risk society (WRS). The remainder of this chapter will explore each primary synergy in turn. Initially, there will be a discussion around the epistemological implications of sustainable development with a focus on the role that science plays in the policy process. This will be followed by exploring the nature of and challenges for governance processes. Here there is an emphasis on the multilateral arena of the UN, followed by a critical analysis of the role of the nation state in sustainable development governance. Finally, there is a brief discussion of the application of this theoretical framework at the local level.

Sustainable Development

Sustainable development is a term that has come to represent the overt acknowledgement within governance frameworks and broader society of the inappropriate development patterns currently in place, as well as a fundamental questioning of the underlying logic of industrial modernity. Sustainable development has been defined as 'Development that meets the needs of the present without compromising the ability of future generations to meet their own needs' (WCED 1987:8). This is the often presented definition of sustainable development coined at the World Commission on Environment and Development in Stockholm in 1987. However, this straightforward and succinct definition obscures a constellation of issues that make the term sustainable development highly ambiguous and contested.

This is not least due to the incompatibility of the two constituent terms of the concept. Development generally refers to a quantifiable enhancement of society through the economic and social well-being of people and society, whereas

DOI: 10.4324/9780429297199-2

sustainability is a subjective and highly qualitative term. Development is a term that is intimately related to the ideas of progress and 'modernisation'. With this in mind, and whilst there is a great deal of research on an interdisciplinary basis on various dimensions of achieving, or perhaps more appropriately 'promoting' sustainable development, there remains a deficit in current research that attempts to access the form that sustainable development takes in governance frameworks.

A number of attempts have been made to categorise and compartmentalise the various elements of the sustainable development literature (Sachs 1999). These assessments of the nature of sustainable development often rely on a framework that outlines strong and weak interpretations of the concept. These interpretations are underpinned by the degree to which sustainable development represents a departure from current modes of production and consumption. Strong sustainability suggests a radical reordering of current socio-political and economic frameworks that are necessary to meet the challenges of current ecological and social crisis (Neumayer 2013; Daly 1997). Whilst there is some evidence that sustainable development manifests in its strongest form, much of the literature points to sustainable development existing in the weaker incarnation. Weak sustainable development operates within the existing socio-political framework; this interpretation of sustainable development focuses on the ability of technology to produce a sustainable future, which is closely aligned with the idea of an ecological modernity.

However, such a polemic model for outlining sustainable development reduces the concept's inherent complexity. Some commentators have attempted to organise what has been identified as fragmented and multidisciplinary position perspectives on sustainable development into a coherent structure designed to guide the observer through the disturbingly muddled literature (Hopwood et al. 2005; Jabareen 2004; Robinson 2004; Redclift 2005).

In light of this ambiguity critics have suggested that conceptually sustainable development is so vague that it can be interpreted as all things to all people and that it is nothing more than well-intentioned fantasy (Norgaard 1994), ultimately making the formulation and implementation of goals difficult because of the many meanings that can arise (Beckerman 1994). Conversely, others argue that the vagueness of the concept is a positive attribute, with ambiguity having the advantage of versatility (Banuri & Najam 2002). Even those who are sceptical of the practical application of the term accept that '…the strength of the idea of sustainable development lies in its ambiguity and its range' (Redclift 1992). Whatever position is taken it is clear that sustainable development is inherently complex and dynamic in nature, and the ways of understanding how sustainable development is being integrated into policy arenas are constantly evolving (Kidd & Fischer 2007; Gazzola 2008).

And this evolution has resulted in the establishment of the United Nations sustainable development goals (SDGs) in 2015, firmly cementing the concept as a strategic vision for humanity's future until 2030 (UN 2025). We will revisit the goals throughout this book. Table 2.1 presents the SDGs with a brief description of each one.

Table 2.1 The Sustainable Development Goals

SDG Number	*Title of SDG*	*Brief Description*
1	No poverty	End poverty in all its forms everywhere.
2	Zero hunger	End hunger, achieve food security, improve nutrition, and promote sustainable agriculture.
3	Good health and well-being	Ensure healthy lives and promote well-being for all at all ages.
4	Quality education	Ensure inclusive and equitable quality education and promote lifelong learning.
5	Gender equality	Achieve gender equality and empower all women and girls.
6	Clean water and sanitation	Ensure availability and sustainable management of water and sanitation for all.
7	Affordable and clean energy	Ensure access to affordable, reliable, sustainable, and modern energy for all.
8	Decent work and economic growth	Promote sustained, inclusive, and sustainable economic growth, full and productive employment.
9	Industry, innovation, and infrastructure	Build resilient infrastructure, promote inclusive and sustainable industrialisation.
10	Reduced inequalities	Reduce inequality within and amongst countries.
11	Sustainable cities and communities	Make cities inclusive, safe, resilient, and sustainable.
12	Responsible consumption and production	Ensure sustainable consumption and production patterns.
13	Climate action	Take urgent action to combat climate change and its impacts.
14	Life below water	Conserve and sustainably use oceans, seas, and marine resources.
15	Life on land	Protect, restore, and promote sustainable use of terrestrial ecosystems.
16	Peace, justice, and strong institutions	Promote peaceful and inclusive societies, provide access to justice, and build strong institutions.
17	Partnerships for the goals	Strengthen global partnerships to support and achieve the SDGs.

Source: UN 2025.

With the establishment of the goals and their subsequent dissemination throughout the global system, the imperative to interrogate the underlying normative assumptions and power dynamics that are associated with the concept is increasingly important. Throughout this chapter we build a perspective on sustainable development that enables us to view the concept in a non-linear and complex way.

Expanding the Risk Society

Risk Society; Towards a New Modernity (Beck 1992) is a sweeping theoretical assessment of humanity's impact on the planet and has achieved a prominent position within the academic community. The publication of the *World Risk Society* (Beck 1999) expanded the scope of the theory and overtly stated the need to understand risk and its consequences from a global perspective. These key texts form the foundation of Beck's thesis and will form the basis for the discussion presented throughout this book. However, as with any good social theorist, Beck has variously elaborated on and adapted the core principles of a risk society, through exploration of globalisation (Beck 2000), terrorism (2002), power (2006), explorations of the normative and political implications of risk, and finally the emancipatory catastrophism of risks (2016). Throughout this book I will refer to a risk society as an umbrella term that incorporates the breadth of Beck's work.

To that end, the underlying message of risk society is that an older industrial society, whose basic principle was the distribution of goods, is being replaced by an emergent 'risk society', structured around the distribution of hazards. For Beck, high-visibility industrial accidents such as those of Chernobyl, Belarus, in 1986 and Bhopal, India, in 1984, as well as large-scale environmental phenomena, such as global warming, force a questioning of the central tenets of modernity (Beck 2016).

Beck explains that risks do not equate to destruction but instead are the probability that destruction from a particular risk may occur. In this context it is the perception of risk that is the driving force for changes in behaviour as well as institutional and political transformation. The risks of a risk society are a manufactured risk created through humanity's impact on the planet and mediated through social mechanisms that Beck refers to as relations of definition. These relations of definition emerge through the key societal institutions of science, government, and the mass media. Within this analysis, Beck distinguishes between three epochs of modernity, which include pre-modernity, industrial or first modernity, and finally, second or late modernity.

In general, the concept of modernity represents a broad set of social relations and processes that typify Western societies. Beck's understanding of modernisation is typical of accepted definitions of this period (Bauman 1998; Giddens 1991). Very broadly modernisation refers to the organisation of social life that emerged from Western Europe. Technological advancements in primary areas such as transportation, communications, and weaponry enabled early nation state entities to expand their influence beyond their own territorial borders to form colonising structures with eventual imperial formations. From the epicentre of Western Europe, modern practices radiated outwards, conflicting with and ultimately subsuming, or juxtaposing, traditional forms of social organisation.

Amongst others, Beck tells us that the processes of modernisation are built on the enlightenment notions of progress and scientific certainty, which hold a dominant position in creating the knowledge upon which progress is founded. During this time, humanity is raised above other species, separating them from the natural global processes to which previous generations of humanity had been bound. Centralised human settlement and intense urbanisation produced new forms of social interaction and innovation. Even in the latter stages of the 19th century, following intense industrialisation, when issues such as environmental despoliation, disease, and hunger were being witnessed on a large scale, science was paradoxically still empowered to solve these problems. From a governance perspective, unitary political structures in the form of nation states aimed to develop policies that jostle for increasing access to, and control of, the world's resources. Moreover, there is increasing success of a capitalist market system which is a driving force of political philosophy in the modern age. Here, the acquisition of wealth is abstracted from its environmental base through the development of a monetary system.

For Beck however, the 21st century's recognition of the different forms of risks caused by untethered scientific development and its inappropriate application by political structures is creating space for another form of modernity. According to Beck, modernity has turned inwards and is questioning its most central tenets, creating a stage of reflexive modernity. Reflexive modernity is a complex collection of processes operating at many different levels. Firstly, Beck argues for macro-analysis of a reflexive modernity by looking at the changing formations of global governance in the face of increasing environmental risk. These are most visible in the changing role of the nation state and the integration of broader social actors into conventional governance structures. Secondly, at the micro level, reflexive modernity exists through altered social relations and the reconstitution of traditional networks of social cohesion. This is occurring both at local governance level and on an individualised basis. Recent work, theoretically and empirically on sustainable development governance, has begun to draw synergies between sustainable development and risk society and particularly with notions of a reflexive modernity (Borne 2018). It has been observed that sustainable development and reflexive modernity provide strong synergies and reinforcing symbiotic perspectives on contemporary human and environmental interaction. The following outlines the primary synergies between sustainable development and the risk society (Table 2.2).

Ultimately, perceiving sustainable development through a risk society lens provides a level of sophistication and an overarching theoretical perspective essential for understanding the intricate and dynamic nature of sustainable development governance. This will lead to an informed assessment of how sustainable development is being integrated in particular governance frameworks and the consequences this has for wider social formations. The following discussion will explore the various elements of sustainable development governance that are highlighted as central from a risk society perspective. The broad synergies identified above

Table 2.2 Synergies between Sustainable Development and a Risk Society

Synergy	*Relationship*
1	Both expose the relationship between humanity and the environment.
2	Draw into question notions of progress, science, and rationality.
3	Emphasise the changing nature of governance structures.
4	Open up the boundaries between the global and the local.

will be used as an organising framework, with each synergy representing a facet of sustainable development governance. The first synergy, exposing the relationship between humanity and nature, was addressed in the previous section and is a constant theme throughout this chapter and this book. The following will therefore begin by exploring the second synergy.

Synergy Two: Progress, Science, and Rationality

The difficulty of interpreting sustainable development begins at the most fundamental level of human endeavour, the creation of knowledge. Drawing on Beck's understanding of a reflexive modernity as a questioning of the central elements of modernistic endeavour, sustainable development is seen from two perspectives. This emerges because, on the one hand, science identifies the environmental problems and the potential risk inherent in them, offering the potential to combat the problems and offer solutions. On the other hand, science and technological advancements, and their unchecked applications, are seen as the cause of these problems in the first instance. Herein lies the initial seed of conflict inherent in the concept. Sustainable development is both a product of science and a stark reminder of its limitations. The consequence of this observation is to illustrate that the production of knowledge through scientific processes is intimately related to wider social processes and the power play that exists therein (Funowicz et al. 1998; Jasanoff 1987, 1990, 2003, 2012, 2021; Renn 2008).

Consequently, the way that sustainable development is applied to a particular risk is fundamentally dependent upon the way that risk(s) are articulated, and the particular lens through which it is viewed fundamentally affects how ensuing visions of the world impinge on the socio-political framework. From an epistemological perspective these visions usually take the form of the dichotomy between realist and constructivist views of the world. The fundamental difference between these two positions is based on the extent to which contemporary risks and indeed reality in general are viewed as culturally constructed or objectively occurring. These perspectives may be summarised as follows: the realist perspective assumes the objective reality of global risks, existing as asocial entities. Constructivism, on the other hand, views the world as mediated completely through the social world, making no distinction between physical existence and interpretation of the external physical phenomena. This constructivist perspective is evident in the works of such commentators as Mary Douglas (Douglas & Wildavsky, 1983), who maintain that risk is a culturally perceived phenomenon, raising questions as to the reality of the current manifestations of risk. But of course over-representing the polemic nature of these positions is hugely reductionist and ultimately misleading. To some degree realists must accept that knowledge is mediated through the social system, and in equal measure constructivists have to accept on some level that reality exists outside the social sphere. To not do so renders both approaches ineffectual as a frame through which to view complex and uncertain phenomena.

To that end an approach is needed that is able to address both institutional and individual actions whilst simultaneously unravelling the motivations for actions and exposing the underlying commitments, or lack thereof, to broader society.

Approaches in this genre have been variously described as contextualised constructivism, co-constructivism, and even crucial realism. Here, I draw on Beck who argues that the question of an either-or is unnecessary. Yes of course we need to identify the framework through which we are viewing the world, but it's important to avoid the ever-decreasing circles of epistemological and ontological debate. Examining our epistemological frame is vital for understanding world views, but so is a willingness to embrace a number of different perspectives, especially when examining discourses of sustainable development. And the way that risk is perceived will fundamentally influence the ways in which adaptive and mitigative policies will be developed. The integration of sustainable development into governance policy will therefore depend very directly on the way that sustainable development is perceived which will be drawn from interpretations of the risk that the concept is employed to address. An extension of this analysis is to highlight how science, risk, and sustainable development relate to policy domains. For example, as Jasanoff (1990) tells us the science-policy advisory process is a fundamentally negotiated and politicised act.

This politicised space has been described by Ravetz (2004) as post-normal science, where uncertainty and the plurality of legitimate perspectives can coexist. Advancing the post-normal model of science, Ravetz (2004) suggests that a more appropriate term for the nature of scientific endeavour is 'precautionary science'. Here, Ravetz draws on some relevant themes, including science in relation to nature, industry, society, and democracy. We can recognise straight away the synergies with other precautionary phrases, notably the precautionary principle, but also precautionary governance, precautionary science, precautionary innovation, and more. So sustainable development doesn't only embody a plethora of risks and perspectives, it also creates a space for a perspective that involves complex interacting systems, and a sustainability science (Kates 2011).

This discussion serves to emphasise the relationships that exist between science, risk, sustainable development, and reflexive modernity. By utilising the WRS the various elements that underpin sustainable development governance and discourse can be exposed. This has included the epistemological discussion on the debate that surrounds constructivism and realism, as well as the interaction between policy and science. These interconnected elements feed directly into how we understand sustainable development governance at both the global and individual levels. Yet these underlying currents remain largely unexplored though the surface textures that they create profoundly impact the future direction of humanity. In what follows, through the structure of the synergies between sustainable development and a reflexive modernity, we can begin to unpick the various domains of sustainable development governance from the global through to the local and individual scales.

Synergy Three: The Changing Nature of Governance Structures

The following section will elaborate on the nature of governance and highlight the complexities that exist in governance structures with a particular emphasis on sustainable development. We will revisit governance as it applies to the local government context in Chapter 3 and then again as it applies at the international level in Chapter 9. For now, we identify governance as a concept that is ambiguous and contested and that in different contexts, governance refers to the many ways that actors across sectors, at diverse scales, interact and negotiate conflicting and diverse priorities. Similarly, Allen et al. (2023) identify governance as the totality of interactions between government, public bodies, as well as the private sector and civil society. Governance understood in this way sits comfortably within a risk society's interpretations of societal interaction because the landscape of a reflexive modernity disrupts traditional political structures.

Applying this analysis to sustainable development, Gupta suggests that this form of governance should be defined as '…the interactive network of regimes at international level, that try and integrate the various elements of sustainable development' (Gupta 2002:361). This definition is indicative of an incorporation of the wider elements of society that now impose on governance frameworks, elements such as corporate pressure and the influence of civil society. This definition however is limited in respect that governance does not just occupy the global level but extends to the local scale. Governance may therefore apply equally to specified local environments and even individual governance in the form of lifestyle change, personal attitude, negotiation of risk, and governance of the self. Earlier observations by Rosenau tell us that identifying the future of governance processes '… is to discern powerful tensions, profound contradictions and perplexing paradox. It is to search for order in disorder, for coherence in contradiction, and for continuity in change' (Rosenau 2005:45). Bearing in mind the levels of ambiguity and complexity that sustainable development represents, Rosenau's insights into the nature of governance provide an indication of why the concept has been so successful in working its way up the political agenda and radiating outwards to all levels of society.

And here we witness the considerable synergies between the processes identified with governance for a sustainable development and those that are articulated within the context of a reflexive modernity. Within the risk society governance, processes themselves become reflexive, becoming adaptive, participatory, and networked. This ultimately leads to a reflexive governance (Beck 1992). Governance must become reflexive in the face of the complex and non-linear issues that are associated with climate change. For Glass and Newig (2019) reflexive governance contributes to the achievement of the SDGs '… by establishing mechanisms that help dealing with the inherent dynamics, uncertainty and complexity related to sustainability problems' (2019:4). This observation provides an important context as the following discussion explores not only multilateral governance but also the establishment of the SDGs on the international stage.

Multilateral Governance

The global aspects of these dynamics will be explored in Chapter 9, but it is valuable to highlight them here to frame the current discussion. There is little doubt that multilateral governance is essential for achieving sustainable development.

Hass (2002) suggests that without well-defined international rules and expectations nation states are unable to unilaterally develop the relevant mechanisms for protecting themselves from global environmental risks. Indeed, sustainable development is unique as a governance guideline by way of its imposition on the global community at the multinational level. In this regard sustainable development has been described as a meta policy (O'Toole 2004). The UN has been a significant proponent of sustainable development, integrating the concept not only into its own operating framework but also into governance structures worldwide. What follows is not an exploration of the functions and operational makeup of the UN; instead what is presented are the main UN events that have galvanised sustainable development on the world stage.

> Global environmental conferences can place new issues on the global agenda and galvanise national concern by publicizing these issues. The conferences often have the effect of reframing issues for decision makers, locating the issues within a new political matrix and thus making possible new tactical and substantive linkages by which policies can be developed.
>
> (Hass 2002:85)

The concept of sustainable development in its modern guise can be said to have developed and been galvanised through the medium of four main events. The first event of considerable significance is the UN Conference on the Human Environment at Stockholm in 1972. The Stockholm conference became a key symbol of political acknowledgement for the growing worldwide awareness of the environment. Whilst this conference produced little in the way of state policy, it achieved two primary successes that act as precursors to the hegemony of sustainable development within the UN and its growth in political discourse. Firstly, it was pivotal in empowering a wider civil society and raising the plight of the environment within the UN system. Secondly, the aftermath of the Stockholm conference resulted in the commissioning of the United Nations Environment Programme (UNEP). The UNEP as well as a number of other UN programmes have been central in creating the global discourse on sustainable development.

The second event towards the advancement of sustainable development was the publication of 'Our Common Future' (WCED 1987). The 1987 World Commission on Environment and Development, popularly known as the Brundtland Commission, provided a common and easily identifiable definition to a previously faceless concept. As already outlined in this book its official definition asserts that sustainable development is development that 'Meets the needs of the present without compromising the ability of future generations to meet their own needs'. The Report 'Our Common Future' brought: '...the discussion of sustainable development to literally millions of people around the world' (Pezzoli 1997:552). Moreover, Irwin (2001) has recognised that 'Our Common Future' represents reconciliation between notions of modernity and environmentalism, and it was this notion of sustainable development that was utilised in the third and certainly the most well-known event, the UN Conference on Environment and Development, or the Rio 'Earth Summit'.

As stated above, the third event to occur and the most publicised was the UN Conference on Environment and Development (UNCED), held in Rio de Janeiro in 1992, popularly named the Earth Summit. It represented a common recognition by many nations and relevant organisations of the world that issues of development needed to be tempered with effective environmental measures. Substantively, a number of agreements emerged from the conference. The Framework Convention on Climate Change set out international targets for reducing the anthropogenic causes of climate change. The Biodiversity Convention established broad aims to conserve international biodiversity, with the aim of making use of components of biodiversity in a suitable manner and enabling an equitable distribution of the benefits of using these resources. The Desertification Convention was designed to create localised action frameworks to address the degradation of dry land environments. Most notable was the production of Agenda 21 (A21), which mapped out guidelines for sustainable development by outlining the main issues implicated in global environmental change and how they might be tackled. The rhetoric contained within this document has had far-reaching implications for the development of sustainable development governance and is the precursor for the form that sustainable development takes today, as represented within the SDGs. The following discussion therefore examines the key elements of A21 and relates these directly to the SDGs.

From A21 to the SDGs

Whilst largely superseded by the SDGs, A21 was a seminal publication and laid the foundations for sustainable development governance at the local level. Its core themes relating to participation and decentralised approaches continue to be relevant for contemporary practice in local governance today. Moreover, A21 set the sustainable development context from within which the research in this book operates. It is therefore educational to elaborate on some of its key features. A21 established a non-legally binding programme of action for sustainable development. It comprised 40 chapters and was intended to guide the actions of governments and non-governmental organisations (NGOs) in issues relating to the environment and development.

A21 has been described as one of the most significant and influential non-binding instruments in the environmental field. For some commentators it was nothing less than the most '…comprehensive document negotiated between governments on the interaction between economic, social and environmental trends at every level of human activity' (Lindner 1997). With that said, and with what will become some familiar and consistent criticisms in relation to sustainable development, it has also been criticised for promoting a vision of sustainable development that does nothing more than perpetuate embedded progress and development, through economic growth and industrialisation. Going even further, other commenters suggested that based on the content of A21, institutional support for sustainable development at the national and international level is actually dysfunctional (Upton 2000). A21 could also be seen to present an overly reductionist role of science in achieving a

Table 2.3 Comparison between SDGs and A21

Feature	*A21 (1992)*	*SDGs (2015)*
Nature	Non-binding framework and action plan	Global strategy with binding political commitment (but still non-legally binding)
Focus	Broad sustainable development principles and recommended actions	Specific, measurable goals and targets (17 goals, 169 targets)
Structure	Thematic chapters with suggested actions (over 800 pages)	Clear set of goals and sub-targets with indicators for monitoring progress
Approach	Flexible guidance; national and local adaptation encouraged	Universally applicable goals for all countries, with monitoring and reporting mechanisms
Monitoring and accountability	Limited, voluntary reporting and informal follow-up	Stronger, regular national reviews, UN High-Level Political Forum oversight
Local engagement	Emphasised through LA21 initiatives	Continued emphasis through localisation of SDGs, but with stronger integration into national policies
Underlying World view	Early sustainable development thinking; influenced by 20th-century environmentalism and participation ideals	Reflexive global modernity; recognition of interconnected risks, climate change, inequality, etc., and need for systemic transformation
Legacy	Foundation for sustainable development discourse; influence on later summits (Rio+20)	Current global blueprint for sustainable development to 2030

sustainable development. However, despite these criticisms, A21 is a pivotal document that deals with a diverse range of sustainable development issues and has had a direct impact on the evolution of the SDGs. Table 2.3 compares the key features of A21 and the SDGs.

A21 enshrined the notion of local action towards sustainable development. The localisation principle is directly visible within the SDG framework, where locating the SDGs is seen as essential for implementation. And whilst the SDGs are interconnected, we especially see this localisation in Goals 11, 13, and 16. And many of the issues identified in A21 map directly to the SDGs. Whilst not definitive, Table 2.4 identifies the key issues in A21 and relates these directly to the SDGs.

The A21 principles were reaffirmed at the World Summit on Sustainable Development (WSSD) (2002) in Johannesburg. There was also more of a focus at that time on practical implementation and partnerships. The global component of research discussed in this book was conducted during the 57th United Nations General Assembly (UNGA), which proceeded the WSSD by only six days. We will

Table 2.4 Do I Need Both of These

A21 Chapter	*Theme*	*Corresponding SDG*
Chapter 3	Combating Poverty	Goal 1: No Poverty
Chapter 4	Changing Consumption Patterns	Goal 12: Responsible Consumption and Production
Chapter 6	Protecting and Promoting Human Health	Goal 3: Good Health and Well-Being
Chapter 7	Promoting Sustainable Human Settlements	Goal 11: Sustainable Cities and Communities
Chapter 8	Integrating Environment and Development in Decision-Making	Goal 17: Partnerships for the Goals
Chapter 9	Protecting the Atmosphere	Goal 13: Climate Action
Chapter 13	Managing Fragile Ecosystems	Goal 15: Life on Land
Chapters 14/15	Promoting Sustainable Agriculture and Managing Biodiversity	Goal 2: Zero Hunger and Goal 15: Life on Land
Chapter 17	Protecting the Oceans, Seas, and Coastal Areas	Goal 14: Life Below Water
Chapter 18	Protecting and Managing Water Resources	Goal 6: Clean Water and Sanitation
Chapter 24	Strengthening the Role of Women	Goal 5: Gender Equality
Chapter 25	Strengthening the Role of Children and Youth	Goal 4: Quality Education
Chapter 33	Financing Sustainable Development	Goal 17: Partnerships for the Goals

revisit this in Chapter 9 and make further connections between sustainable development and reflexive modernity, but it is pertinent to note the proximity of these events as impacting the discourses of sustainable development observed.

Ten years later at Rio+20 in 2012 there was general recognition that a new measurable framework was needed, and the outcome document 'The Future we Want' called for the establishment of the SDGs that replaced the Millennium Development Goals (MDGs) in 2015.

In line with the previous discussion exploring the transition from A21 to the SDGs, Lidia Brito (2012), Director of UNESCO, indicated that the SDGs differ from MDGs in three main ways. First, MDGs are centred around seven social goals and just one environmental goal. Long-term social and economic improvement will need closer attention to be paid to the environment. Second, whereas MDGs focused on developing countries, SDGs need buy-in from all nations. Third, MDGs were hastily assembled without thorough analysis (see Stafford-Smith et al. 2017). Subsequently, the United Nations Sustainable Development Summit adopted the 17 SDGs and the 169 targets that are associated with them.

Whilst these events chart the evolution of sustainable development, they are as open to criticism as is the concept itself. They have been described as fleeting moments of media frenzy. Death (2011) has argued that these events 'reinforce dominant hierarchical, state centric, elitist and rationalist model of politics and have symbolic, performance and theatrical roles' (2011:2). Swyngedouw (2010) suggests that these events fail to provide real possibilities for different socio-environmental futures. And whilst, as with consistent criticisms associated with sustainable development, I do not disagree with these assessments, I do gravitate towards perspectives that suggest it is not useful to define the success or failure of such events by the number of action plans and governmental initiatives that are produced (Buckingham Hatfield and Walker 2002). Instead success should be gauged by the extent to which there is fresh impetus for action, personally and collectively, as well as reinforcing wider themes such as socialisation and education (Hass 2002). Moreover, it is recognised that visible international events are essential in establishing transnational 'discourse coalitions', capable of binding opinion on environmental issues and fundamentally influencing international relations (Hajer 2002; Dryzek & Stevenson 2011). Extending this analysis further the following discussion introduces Synergy Four which highlights the connection between the global and the local.

Synergy Four: Global and the Local

Local Sustainable Development Governance

Initially, and by way of connecting the global and local realms of sustainable development, it is pertinent at this stage to draw again on A21, which, as we have seen, cascades from the global level to the local and individual levels of analysis as well as mapping directly onto the current SDGs. By doing so we can chart the evolution of sustainable development governance at the local level and relate this directly to local components of the SDGs. Chapter 28 of A21 focuses directly on the role of local communities in achieving a sustainable development, calling for each community to formulate its own governance initiatives (Agyeman & Angus 2003). These processes have been encompassed by the more specific Local Agenda 21 (LA21). In essence, LA21 proposes that local authorities, in conjunction with local residents and NGOs, should produce a strategy designed to further a sustainable development (Chatterton & Style 2001).

The emphasis is on creating a multi-stakeholder environment in order to provide the opportunity for new forms of dialogue within governance structures. This is a dialogue that highlights the possibility of inclusionary political processes, incorporating previously marginalised sectors of society, whilst at the same time attempting to highlight the structural limitations to the impact of locally based sustainable development programmes that may already be in operation (Marvin & Guy 1997). To a certain degree LA21 can be seen as a departure from the top-down prescriptive governance structures of a first modernity. This is indicative of a reflexive governance structure, formulated on the basic tenets of achieving a sustainable development. With this said however, there has been some criticism of the inherent functionality of practices put in place through LA21 initiatives (Worthington et al. 2003). For example, the practical outcomes of an opening up of the governance structure through the establishment of LA21 have been criticised for failing to manifest in the long term (Lafferty & Eckerberg 2013; Evans & Theobald 2003).

However, building on the premises identified in LA21 the UK's national sustainable development strategy 'Securing the Future: Delivering the UK Sustainable Development Programme' continued this rhetoric by emphasising the importance of local communities in enhancing sustainable development. The coalition government's Big Society project and the Localism Act (2011) provided emerging areas for debate concerning local governance and sustainable development. Though it is recognised that these too have been critiqued for inconsistencies between rhetoric and practice (Clarke & Newman 2012). These observations can be extended to recent policy and legislative areas, and these will be critically discussed in more detail in Chapter 3.

Globalisation and the Role of the Nation State

The final entity that needs to be discussed within the context of Synergy Three is the nation state. The nation state has traditionally been understood as an ensemble of coercive, administrative, and legal institutions that claim sovereignty and predominance over other organisations within their boundaries (Azarya 1988). This view persists today where it is recognised that the state is an autonomous institutional complex, reinforcing rules and exercising sovereignty and control over its territory (Hay et al. 2022). However, this first modern interpretation of the nation state is criticised in an age of increased interconnectivity and globalisation. Beck cautions against proclaiming the end of the nation state, which in many regards, and in certain instances, has become more powerful, but nevertheless has suggested that in a globalised WRS the nation state has become a zombie category. In this context Beck maintains that the nation state is a category of industrial/first modernity and no longer applies in an age of global risk. The state is then reimagined as a fluid grouping of institutions with unstable boundaries. Fluid because it must cooperate internationally, through multiple mechanisms, climate agreements, trade deals, and as such shape power with supranational organisations such as the UN, the European Union, and the World Trade Organization.

There is an emphasis on the permeability and flexibility of boundaries, with contemporary societies existing both above and below the state system. In this context power flows are decentralised and networked. Networks of communication, finance, migration, activism, and more bypass the traditional nation state control. Whilst it's recognised that societies in the modern world have always operated in complex relationships that function beyond the nation state system, it is the increasingly globalised nature of the world that has taken this to another level. In particular, the globalisation of environmental phenomena, such as global warming, which have given rise to the need for the idea of sustainable development, permeates state boundaries, undermining the sovereignty of individual states.

However, as Frickel and Davidson (2004), drawing on Habermas (1987), recognise in relation to sustainable development, nation states are responsible for the ratification of relevant international conventions and agreements, the adoption of necessary legislation, the enactment of new laws, and the harmonisation of existing laws. Through processes of rationalism and legitimisation, Frickel and Davidson

Table 2.5 Nation State in a Reflexive Modern Era

Period	*Key Features*	*Transition Drivers*
Traditional modernity (18th–20th century)	Nation state as unitary sovereign: Strong borders Centralised power National economic management Monopoly of force	
Transition		Industrialisation, risk creation, global interconnection
Risk society/reflexive modernity (late 20th century onwards)	Nation state as fluid and contested: Global risks (climate change) Diffusion of authority Need for global cooperation Public reflexivity and uncertainty	
Transition		Globalisation, digital networks, flow of capital and people
Networked global society (contemporary world)	Nation state embedded in global networks: Flexible, negotiated sovereignty Governance through multilevel systems Dynamic management of transnational risks	

suggest that the rationalisation of sustainability within the state forms a new political dialogue that redefines ecological conditions to emphasise risk and crisis. This new political dialogue may be seen as emerging through the discourse of sustainable development as states attempt to manage the environmental, social, and economic crisis both within and beyond their boundaries. In this context states shape sustainable development not merely through compliance, but by integrating legitimacy and rational governance processes into their environmental policy-making (Rea & Frickel 2024). Positioning this within a reflexive modernity, Table 2.5 outlines the evolution of the nation state in a reflexive modernity.

The integration of sustainable development principles into the governance frameworks of states, which resulted directly from the Rio conference and A21, manifested in the form of national sustainable development strategies. These were participatory and evolving sets of analysis '…which integrates the economic-social and environmental objectives of society, seeking trade offs where this is not possible' (Dalal-Clayton & Bass 2002:31). National Sustainable Development Strategies moved away from a model that focuses on central planning, to one that is capable of creating enabling conditions that should be based around improving and making strategic connections between existing strategic planning frameworks. Guiry (2024) has further developed the analysis in relation to the SDGs asking why states adhere to the SDGs. Exploring elements of adherence and legitimacy, Guiry concludes that the '… near-universal acceptance of and reporting on a non-binding framework by states has created a newfound global paradigm, the normative impacts of which are yet to be fully established' (2024:426). Reinforcing previous comments in this chapter, utilising a risk society and exploring the potential of an emergence of a reflexive modernity directly addresses the normative underpinnings of sustainable development and by extension the SDGs.

Conclusion

This chapter's aim has been to draw essential and fundamental connections between sustainable development and a WRS. The goal has not been to provide a prescriptive one-size-fits-all framework for exploring sustainable development. Instead, by examining the synergies between sustainable development and the WRS, a complex array of issues has been tied together at multiple scales.

Initially, sustainable development was presented as an innately multifaceted and contested term. This was followed by outlining the main elements of Beck's Risk Society thesis. The synergies that exist between the sustainable development and a risk society were then outlined, and these synergies were used to guide the discussion from the global to the local scales of analysis. Following this the conflicting roles that science plays in society were explored. The international governance scale was explored through a discussion on the importance of multilateral cooperation in achieving sustainable development, with an emphasis on the UN. This was particularly emphasised by using the main events that have helped galvanise sustainable development on the international stage. The discussion then moved to the national level, where the fluid and interactive nature of the nation state was discussed in conjunction with observations on the risk society. Overall this chapter has argued that in order to move forward with the sustainable development agenda a realistic interpretation of the complex processes that are involved needs to be translated into constructive policy initiatives. The relationship between sustainable development and a risk society is an evolving agenda, but the continued debates that exist between theory and empirical observation enable researchers and policy-makers to move beyond the often dogmatic and reductionist rhetoric that surrounds sustainable development governance.

Bibliography

Allen, C., Malekpour, S., & Mintrom, M. (2023). Cross-scale, cross-level and multi-actor governance of transformations toward the sustainable development goals: A review of common challenges and solutions. *Sustainable Development*, *31*(3), 1250–1267.

Agyeman, J., & Angus, B. (2003). The role of civic environmentalism in the pursuit of sustainable communities. *Journal of environmental Planning and Management*, *46*(3), 345–363.

Agyeman, J., & Evans, J. (2004). Just sustainability: The emerging discourse of environmental justice in Britain? *Geographical Journal*, *170*, 155–164.

Azarya, V. (1988). Reordering state-society relations: Incorporation and disengagement. In D. Rothchild & N. Chazan (Eds.), *The precarious balance: State and society in Africa* (pp. 10, 3–21). Westview Press.

Banuri, T., & Najam, A. (2002). *Civic entrepreneurship; a civil society perspective on sustainable development*. Volume 1, A global synthesis. Grandhara Academy Press.

Bauman, Z. (1998). *Globalisation, the human consequences*. Polity Press.
Beck, U. (1992). *Risk society: Towards a new modernity* (M. Ritter, Trans.). Sage. (Original work published 1986).
Beck, U. (1994). *Ecological politics in an age of risk*. Polity Press.
Beck, U., Giddens, A., & Lash, S. (1994). *Reflexive modernization: Politics, tradition and aesthetics in the modern social order*. Polity Press.
Beck, U. (1999). *World risk society*. Polity Press.
Beck, U. (2000). *What is globalization?* Polity Press.
Beck, U. (2002). *The terrorist threat: World risk society revisited*. Polity Press.
Beck, U. (2006). *Power in the global age: A new global political economy*. Polity Press.
Beck, U. (2009). *World at risk*. Polity Press.
Beck, U. (2016). *The metamorphosis of the world*. Polity Press.
Beckerman, W. (1994). Sustainable development: Is it a useful concept? *Environmental Values*, *3*(3), 191–209.
Borne, G. (2018). *Surfing and sustainability*. Routledge.
Brito, L. (2012). Analyzing sustainable development goals. *Science*, *336*(6087), 1396–1396.
Buckingham-Hatfield S., and Walker G. (2002). 'From UNCED to the WSSD: Reasons to be Cheerful'. *Local Environment*, *7*(3), 237–240.
Buckingham-Hatfield, S., & Matthews, J. (1999). Including women: Addressing gender. In S. Buckingham-Hatfield & S. Percy (Eds.), *Constructing local environmental agendas*. Routledge.
Chatterton, P., & Style, S. (2001). Putting sustainable development into practice? The role of local policy partnership networks. *Local Environment*, *6*(4), 439–452.
Clarke, J., & Newman, J. (2012). The alchemy of austerity. *Critical Social Policy*, *32*(3), 299–319.
Dalal-Clayton, B., & Bass, S. (2002). *Sustainable development strategies a resource book*. International Institute of Environment and Development.
Daly, H. E. (1997). *Beyond growth: The economics of sustainable development*. Beacon press.
Death, C. (2011). Summit theatre: Exemplary governmentality and environmental diplomacy in Johannesburg and Copenhagen. *Environmental Politics*, *20*, 1–19.
Douglas, M., & Wildavsky, A. (1983). *Risk and culture*. University of California Press.
Dryzek, J. S., & Stevenson, H. (2011). Global democracy and earth system governance. *Ecological Economics*, *70*(11), 1865–1874.
Evans, B., & Theobald, K. (2003). Local Agenda 21 and the shift to 'governance' in the UK. In W. M. Lafferty (Ed.), *Implementing LA21 in Europe* (pp. 111–132). Earthscan.
Frickel S., and Davidson D. (2004). Building Environmental States: Legitimacy and Rationalisation in Sustainability Governance. *International Sociology*, *19*(1), 89–110.
Funowicz, S., Ravetz, J., and O'Conner, M. (1998). Challenges in the use of science for sustainable development. *International Journal of Sustainable Development*, *1*(1), 99–107.
Gazzola, P., (2008). What appears to make SEA effective in different planning systems. *Journal of Environmental Assessment Policy and Management*, *10*(1), 1–24.
Giddens, A. (1991). *Modernity and self identity*. Polity Press.
Glass, L. M., & Newig, J. (2019). Governance for achieving the sustainable development goals: How important are participation, policy coherence, reflexivity, adaptation and democratic institutions? *Earth System Governance*, *2*, 1–14.
Guiry, N. (2024). Why do states adhere to the sustainable development goals? *Transnational Environmental Law*, *13*(2), 400–427.

Gupta, J. (2002). Global sustainable development governance: Institutional challenges for a theoretical perspective. *International Environmental Agreements: Politics, Law and Economics*, *2*(4), 361–388.

Habermas, J. (1987). *The theory of communicative action, Vol ii*, trans. T. Maccarthy. Cambridge, Polity Press.

Hajer, M. A. (2002). Discourse coalitions and the institutionalization of practice: The case of acid rain in Great Britain. In *Argument turn policy anal plan* (pp. 51–84). Routledge. https://www.researchgate.net/publication/238218883_Discourse_Coalitions_and_the_Institutionalization_of_Practice_The_Case_of_Acid_Rain_in_Great_Britai

Hass, P. (2002). United Nations conferences and the constructivist governance of the environment. *Global Governance*, *8*(1), 73–91.

Hay, C., Lister, M., & Marsh, D. (2022). *The state: Theories and issues*. Bloomsbury Publishing.

Hopwood, B., Mellor, M., & O'Brien, G. (2005). Sustainable development: Mapping different approaches. *Sustainable Development*, *13*(1), 38–52.

Irwin, A. (2001). *Sociology and the environment: A critical introduction to society, nature and knowledge*. Polity Press.

Jabareen, Y. (2004). A knowledge map for describing variegated and conflict domains of sustainable development. *Journal of Environmental Planning and Management*, *47*(4), 623–642.

Jasanoff, S. (1987). Contested boundaries in policy relevant science. *Social Studies of Science*, *17*, 195–213.

Jasanoff, S. (1990). *The fifth branch: Science advisors as policy makers*. Harvard University Press.

Jasanoff, S. (2003). Breaking the waves in science studies: Comment on H. M. Collins and Robert Evens, the third wave of science studies. *Social Studies of Science*, *33*(3), 389–400.

Jasanoff, S. (2012). *Science and public reason*. Routledge.

Jasanoff, S. (2021). Knowledge for a just climate. *Climatic Change*, *169*(3), 36.

Kates, R. (2011). What kind of science is sustainability science? *PNAS*, *108*(49), 19449–19450.

Kidd, S., & Fischer, T. B. (2007). Towards sustainability: Is integrated appraisal a step in the right direction? *Environment and Planning C: Government and Policy*, *25*, 233–249.

Lafferty, W., & Eckerberg, K. (Eds.). (1998). *From the Earth Summit to Local Agenda 21: Working towards sustainable development*. Earthscan.

Lafferty, W. M., & Eckerberg, K. (Eds.). (2013). *From the Earth Summit to Local Agenda 21: Working towards sustainable development*. Routledge.

Lindner, C. (1997). Agenda 21. In F. Dodds (Ed.), *The way forward: Beyond Agenda 21* (pp. 1–12). Earthscan.

Marvin, S., & Guy, S. (1997). Creating myths rather than sustainability: The transitions fallacies of the new localism. *Local Environment*, *2*(3), 299–302.

Naess, A. (1973). *The shallow and the deep, long range ecology movement: A summary*. Scandinavian University Press.

Neumayer, E. (2013). *Weak versus strong sustainability: Exploring the limits of two opposing paradigms*. Edward Elgar Publishing.

Norgaard, R. (1994). *Development betrayed: The end of progress and the co-evolutionary revisioning of the future*. Routledge.

O'Toole, L. (2004). Implementation theory and the challenge of sustainable development: The transformative role of learning. In M. Lafferty (Eds.), *Governance for sustainable development: The challenge of adapting form to function* (pp. 32–60). Edward Elgar.
Pezzoli, K. (1997). Sustainable development: A trans-disciplinary overview of the literature. *Journal of Environmental Planning and Management*, *40*(5), 549–574.
Ravetz, J. (2004). The post normal science of precaution. *Futures*, *36*, 347–357.
Rea, C. M., & Frickel, S. (2024). Environmental state. In C. Overdevest (Ed.), *Elgar encyclopedia of environmental sociology* (pp. 270–279). Edward Elgar Publishing.
Redclift, M. (1987). *Sustainable development: Exploring the contradictions.* Methuen.
Redclift, M. (1992). Sustainable development: Needs, values, rights. *Environmental Values*, *2*(1), 3–20.
Redclift, M. (2005). Sustainable development (1987–2005): An oxymoron comes of age? *Sustainable Development*, *13*(4), 212–227.
Renn, O. (2008). *Risk governance, coping with uncertainty in an uncertain world.* Earthscan.
Robinson, J. (2004). Squaring the circle? Some thoughts on the idea of sustainable development. *Ecological Economics*, *48*(4), 369–384.
Rosenau, J. (2005). Governance in the 21st century. In R. Wilkinson (Ed.), *The global governance reader*. Routledge.
Stevens, C. (2023). Strengthening reflexive governance to achieve the 2030 Agenda and the SDGs. *International Environmental Agreements: Politics, Law and Economics*, *23*(2), 173–178.
Swyngedouw, E. (2010). Apocalypse forever? Post-political populism and the spectre of climate change. *Theory, Culture, Society*, *27*, 213–232.
Upton, S. (2000). Roadblocks to Agenda 21. In F. Dodds (Ed.), *Earth summit 2002* (PP. 3–20). Earthscan.
United Nations, Department of Economic and Social Affairs. (2025). The Sustainable Development Goals Report 2025. United Nations. https://unstats.un.org/sdgs/report/2025/
World Commission on Environment and Development (WCED). (1987). *Our common future*. Oxford University Press.
Worthington, I., Patton, D., & Lindley, I. (2003). Local authorities, business and LA21: A study of the East Midlands sustainable development partnerships. *Local Government Studies*, *29*(1), 91–110.

3 Local Governance in a Global Context

As we have already seen the role of these local-level bodies remains underexplored and contested but has benefited in recent years from increased theoretical, empirical, and policy scrutiny (Wills 2016; Willett 2021). Even though town and parish councils directly engage with communities, planning, conservation, local infrastructure, and more, they are often marginalised from broader governance debates on sustainable development and associated areas such as global warming, economic development, and environmental management (Creamer et al. 2019a, 2019b). This chapter therefore explores town and parish councils within identified sustainable development contexts, examining how councils navigate sustainability governance within England's multilevel institutional landscape, assessing their role in areas such as local planning and land-use governance under the Localism Act 2011, climate action and biodiversity management under the Environment Act 2021, community-led renewable energy projects, as well as net-zero transitions, and social and economic sustainability initiatives at the grassroots level.

This analysis adopts a sustainable development lens which identifies key opportunities, contradictions, and systemic constraints that shape how town and parish councils engage with broader sustainable development processes. By doing this the chapter not only contextualises local councils but also addresses a critical gap in governance and policy research, offering insights into the structural challenges facing local sustainable development governance. Moreover, issues discussed in this chapter inform, and are informed by, the discussion and presentation of data in subsequent chapters.

In Chapter 2 we explored the relationship between sustainable development and a risk society, and we contextualised this within governance processes at the international and local levels. At the local level we can elaborate further to identify the sub-political space that local government operates within. To that end we now identify town and parish councils within a risk society as occupying a sub-political space. I have previously drawn on the notion of sub-politics as a conceptual space that refers to political formations that sit outside traditional party political structures. In the context of my precious research, the notion of politics was applied to a project designed to reduce energy consumption in the home, this Carbon Reduction Innovation Support Pilot, or CRISP for short. The programme was run as a partnership between Plymouth City Council and the charity Global Action Plan

DOI: 10.4324/9780429297199-3

for the Earth. Earlier interpretations of Beck's sub-political space identified a complete detachment from the conventional political base of the nation state, bypassing political parties and representative opinion.

In this book, we identify a more sophisticated and nuanced interpretation of sub-politics, and it is important to set the parameters of the concept at this point. As a starting point, Beck argues that sub-politics is characterised by the individualisation of responsibility, which is reinforced through a decentralisation of authority, which is an inevitable social reflexivity that is a response to global risk. For Beck, sub-politics represents individual freedom from traditional social structures, such as family, class, religion, and conventional political processes. But in reality the complete separation of the political and sub-political is reductionist and not representative of the reality of complex and interrelated social structures. As Mythen (2004) recognises the separation of political and sub-political is an illusion, and Beck appears not to acknowledge that discourses of risk are embedded in institutional mechanisms for regulation as well as managing opposition that threatens the traditional power base. For example, if we turn to Foucault for a moment we understand that social institutions are responsible for constructing an understanding of risk, intimately affecting as well as restricting behaviour. This is the underlying premise of Foucault's governmentality, where it is suggested that '…expert institutions employ discourses of risk that maintain normative parameters' (Borne 2006:89). Similarly, we can briefly call upon Gramsci to inform the relationship between political elites and wider society as a combination of coercion and consent. Where Beck sees sub-politics as freedom from conventional political norms, Gramsci tells us that dominant groups operate not only through coercive mechanisms, legal and political systems, but also through processes that facilitate consent, mediated through culture, media, religion, and established social norms. So whilst Beck sees sub-politics as opening up emancipatory spaces where individuals and groups can shape their own governance processes, Gramsci reminds us that established power can permeate these spaces and influence their dynamics. We may then witness a scenario where these spaces do not serve to bypass or disrupt established political norms, but instead reinforce them.

With that in mind Beck and Gramsci are not at polemic opposites when it comes to interpretations of political and counter-political forces. Gramscian notions of counter-hegemony, where groups can challenge dominant risk framings, expose contradictions, and create alternative narratives and pathways through the exploration of political and social theory. With reference to sustainable development and the exploration of the emergence of a reflexive modernity, we remain vigilant in light of these insights. These perspectives allow us to constantly challenge the dominant narratives of sustainable development, identifying instances where these narratives may represent the extension of institutional power without fundamental systemic change. And equally these discussions help us identify narratives where sustainable development is presenting something new, something that does represent a truly reflexive modernity. The research presented in this book is not only sensitive to this power imbalance but also seeks, through in-depth discourse analysis, to expose these processes as an essential part of the construction of the narrative of sustainable development.

With the formulation of sub-politics established we now move to conceptualise towns and councils as a nexus point, an assemblage of multiple networks and flows that converge on this sub-political space, and then are redirected outwards through formal and informal governance processes. This interpretation of the parish council space represents an extension of the way that the local-level analysis of sustainable development and reflexive modernity coalesced through a single case study. Research presented in this book takes those initial insights forward and applies them in a more comprehensive and sophisticated way, moving beyond a single case study to explore an entire level of local government that is embedded in the assemblages of their local communities.

Indeed the notion of assemblages emerges as a key metaphor for interpreting the permeability of the spaces under investigation. In Chapter 2 we explored knowledge formation, identifying hybrid structures, nexi, and assemblages as elements of a reflexive modernity. This interpretation can be extended again to the spaces of local government where multiple processes converge.

Willett (2021) demonstrates that research that utilises assemblages has a rich history both within and beyond the social sciences. The use of assemblages as a conceptual frame, '…enables a rich and textured analysis of the deep complexity of the ways that people, ideas, and institutions interact, connect, break apart, and recombine with regard to the specific environments they are located in' (2022:4). Again as Willett suggests, the advantage of this approach is to explore the becoming of local areas and view them as an interactive and dynamic space. This approach also addresses recognised limitations within border local government studies literature as to how the idea of 'local' is conceptualised. As Giovannini and Griggs (2024) point out, whilst there is a critical understanding of the space of 'the local', what constitutes local in the context of local government has become the elephant in the room. The authors argue that '… the real value of the local both as a concept and practice, lies in its fuzzy often problematic nature' (2024:989). There is also recognition that rise of governance narratives and processes have blurred borders, responsibility, and accountability. Here I suggest that assemblies within the political framework both invigorate the local space of activities, whilst recognising the interconnected dynamics that inevitably impact upon it.

The discussion to date has further developed and contextualised the space within which the empirical work was conducted. The exploration of sub-politics helps us critically understand the dynamism of the local government space, as well as focusing more directly on the relationship between sustainable development and reflexive modernity. The rest of this chapter seeks to connect these broader theoretical observations to tangible governance structures and processes that impact town and parish councils. The form of governance is initially identified within the tradition of participatory governance, which is then connected to the risk society. Multi-level governance, social capital, and technological progress further capture specific dynamics for the local government arena. This is followed by a historical overview of the emergence of parish councils as set within wider political paradigms, and this is ultimately related to the contemporary context. Finally this chapter identifies key legislative and policy initiatives that both enable and constrain local government from a sustainable development perspective.

Participatory Governance

Initially it is recognised that within the context of sub-politics, I use participatory and deliberate governance interchangeably. Although these concepts have different theoretical and academic backgrounds, the practice and intentions of each frequently converge. As we discuss sustainable development within local governance and relate this to the global scale, this pragmatic approach reflects the hybrid nature of political engagement within a risk society (See Bua & Bussu 2023).

As we have already identified, traditional politics through government processes of parliament and political parties are no longer the only or even the most desirable arenas for decisions to be made at the local level. Sub-politics identifies new forms of political assemblages, such as citizen groups, NGOs, businesses, and extended local networks across sectors. Within this context we are able to identify participatory governance as a formalisation or institutionalisation of sub-political processes. This creates a space for intentionally including non-traditional sectors in sustainable development and risk-related decision-making with effective action involving actors beyond formal government. However, as we have already alluded, this raises many questions as it addresses. Is the participation we are discussing meaningful and effective, or is it simply tokenistic? How do we identify who actually gets to sit at the table in the first place and participate? And does participatory governance subdue or empower sub-political forces? Figure 3.1 outlines the relationship of participatory governance in a risk society.

Theories of deliberative democracy suggest that effective governance requires inclusive, informed, and equitable public participation (Dryzek 2002). In relation to sustainable development, participatory models emphasise elements such as citizen-led decision-making rather than top-down governance, social learning, and consensus-building through public deliberation and the co-production of policies between councils and communities. As the closest level of government to local communities, it seems sensible to suggest that town and parish councils are well placed to enact the principles of participatory governance. However, building on the issues we have already identified in this chapter, engagement with town and parish councils remains unequal, inconsistent, and vulnerable to what has been termed elite capture. As Parker et al. (2015) observe, '… the challenge becomes how to design and implement engagement that is somehow proofed against the problems of instrumentalism, co-option and colonisation by more dominant actors and interests' (2015:2).

These observations feed directly into concepts of social capital which depend upon networks of trust and norms that enable collective action (Putnam 2000; Gelderblom 2018). As Skerratt and Steiner (2013) point out, low social capital communities struggle with participation fatigue, fragmentation, and disengagement from local governance (also see Steiner & Farmer 2018). These observations are tempered with Watkins' (2021) observations on social capital that take a critical approach to the term and its capacity to represent community capacity.

Set against this discussion, the capacity of town and parish councils to engage in any meaningful way in embedding sustainable development within their local communities is therefore deeply embedded in social relationships as well as historical

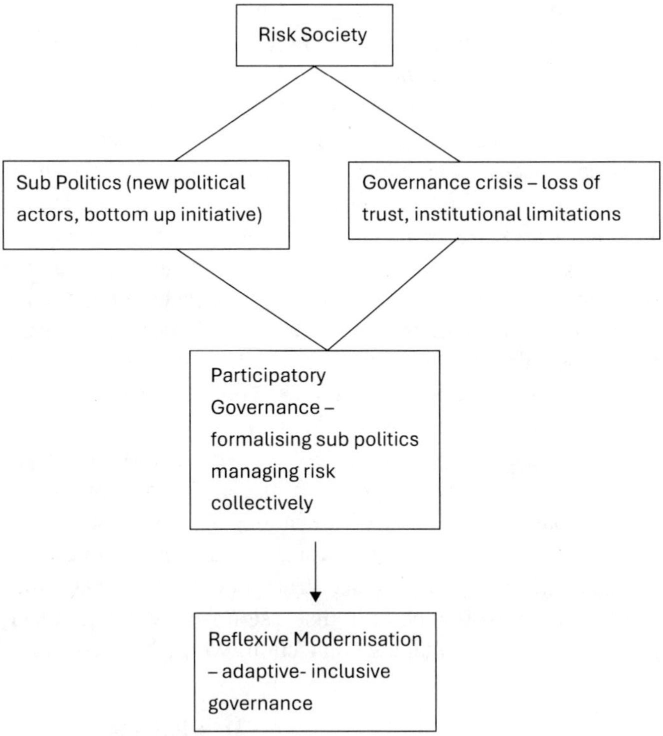

Figure 3.1 Participatory Governance in a Risk Society

patterns of civic participation. This form of participation is achieved in a number of ways and through the mediums of particular participatory mechanisms.

With issues relating to elite capture in mind (Bussu et al. 2022), councils can hold public forums where residents are able to discuss priorities and sustainable development-related issues. Some councils have also introduced participatory budgeting, allowing citizens to allocate funding towards local sustainability projects (Sintomer et al. 2016). Processes relating to co-production and collaboration between councils, residents, and local organisations present opportunities to address locally relevant priorities that are inclusive and meaningful to those local areas (Brandsen & Pestoff 2006). Initiatives might include community-managed green spaces, local food, and agroecology projects, volunteer-led biodiversity conservation projects, and much more. Whilst opportunities exist it is worth recognising that the co-production process can be time and resource intensive, requiring enduring commitments from all parties involved (Lowndes & Sullivan 2004). It is also important to recognise the potential impact that the continued advancement and integration of technology is having on these processes.

Notably the rapid onset of digitalisation and digital governance tools and artificial intelligence platforms provides enhanced opportunities to expand community engagement. Town and parish councils increasingly use social media, e-petitions, and digital consultations to engage residents. This observation is set within the context that there is uneven engagement with digital technologies, especially amongst rural communities. In recent years Cornwall has benefited from investments such as Superfast Cornwall, expanding high-speed broadband across rural areas. However, concerns remain over digital exclusion. Over 90 per cent of households in Cornwall now have access to broadband, but actual adoption rates and meaningful use are uneven (Ofcom 2023). More broadly it has been recognised that the smaller councils may not possess the technical capacity to implement digital engagement strategies effectively. This is an ongoing discussion and an important area for continued investment and research (LGA 2025). These inequalities are further compounded by persistent demographic inequalities, such as age. For example, older generations may be more inclined to engage with formal governance structures that include town and parish councils. In contrast younger generations lean towards informal grassroots activism (Fox et al. 2012).

Whilst there is significantly more to say on participatory governance and the issues raised, it is the goal of the previous discussion to draw out some of the opportunities and barriers for creating an enabling environment for local councils to embed sustainable development in local communities. In so doing we take a broad brush approach to the theoretical, structural, and practical issues that coalesce around this topic. At this point in discussion we are going to step back and expand the scope of analysis, making connections with other tiers of government.

Multilevel Governance and Local Sustainable Development

One of the central challenges in sustainable development and associated complex phenomena is ensuring coherence between different levels of governance (Jordan & Huitema 2023). Indeed the foundations upon which this book is built have explored exactly this issue as the narrative moves from the international level of the United Nations to local government and local communities. As we have already discussed in this chapter town and parish councils are embedded within both horizontal and vertical governance dynamics, including national and devolved governance entities. This can include regional and local government statutory bodies, district and county councils, as well as non-governmental, third sector, and business organisations. All of the aforementioned shape both the discourses of sustainable development and the ground outcomes through the mobilisation of public opinion and resources, as well as civic engagement and private investment.

These multilevel governance dynamics inevitably provide both opportunities and barriers for achieving a sustainable development at the local level. Whilst decentralisation rhetoric, policy, and legislation may be seen to strengthen the role of town and parish councils, in reality, this layer of government in regard to funding, capacity, and legal authority remains significantly constrained, leading to fragmented and uneven engagement with sustainable development implementation (Parker et al. 2023). Inevitably then, policy responsibilities are often poorly defined and inconsistent, resulting in widely recognised governance inefficiencies. As Copus et al. (2017) point out English local government has been shaped by a persistent tension between centralist control and localist rhetoric, which ultimately creates structural constraints that limit local autonomy. The historical evolution of this tier of government provides further insights into contemporary capacity and sustainable development integration.

Historical Context and Evolution

Town and parish councils have their origins in the Local Government Act 1894, which established them as the first tier of local government (see Jones 2020). Wills (2016) provides an extensive and in-depth exploration of the development of parish councils within the context of localism, highlighting that over time, parish councils' structure, context, and responsibilities have significantly evolved. Wills detailed the way that '... English local government shifted from judicial regimes embedded in the legal and ecclesiastical infrastructure provided by the crown and the church, towards one more focused on the power of national parliament' (Wills 2016:71). Here, Wills makes the connection with Foucault's governmentality where the state is increasingly focused on '...the welfare of the population, the improvement of its condition, the increase of its wealth, longevity and heath' (Wills 2016:71). We can also connect this observation to those of the exploration of the nation state in Chapter 2 as a central theme in the relationship between sustainable development and a reflexive modernity.

Ultimately, the ability of parish councils to embed sustainable development at the local level is dependent on structural limitations as they continue to operate within a governance framework, where they have discretionary rather than statutory responsibilities in most policy areas, and they lack direct enforcement mechanisms for environmental and planning decisions. Parish councils' financial autonomy is limited, with most funding derived from the parish precept, grants, or voluntary contributions. These constraints shape how sustainable development is implemented at the local level, often leading to highly uneven patterns of impact and effectiveness.

Given the complex governance environment within which town and parish councils operate, this chapter seeks to set a context for local government, providing a foundation for the subsequent presentations and analysis of empirical data. To that end this chapter asks the following questions: To what extent do town and parish councils influence sustainable development policy and practice at the local level? What are the primary barriers preventing them from implementing effective sustainability policies? How do governance processes, financial constraints, and policy contradictions shape local sustainability initiatives? These questions will then inform subsequent discussions on what structural reforms may be required to enhance the role of town and parish councils within a sustainable development context.

Parish Councils and the Cornish Context

It is clear by now that the central rationale for the research presented in this book is to explore narratives of sustainable development and issues associated with the concept. But whilst at the time of data collection the concept was on the rise in political, social, and environmental discourses, it did not enjoy the lofty position it now occupies. As we will see in subsequent chapters mechanisms for understandings and embedding sustainable development were disparate, and ultimately it was necessary for me to draw these disparate mechanisms together under the banner of sustainable development in order to explore relevant narratives.

Increasingly, the challenges and opportunities for embedding sustainable development in local government are being more directly addressed with frameworks being explored that can accommodate complex system dynamics. For example, in Cornwall, sustainable development has been explored within the context of the so-called doughnut economics framework (Turner & Wills 2022; Turner et al. 2021). And as Turner et al. (2021) illustrate, there is a need to identify indicators that '…enable longer-term monitoring of social and environmental trends… support the identification of priority areas for action on sustainability challenges'.

And these frameworks are explicitly and actively incorporated into strategic visions for Cornwall's Future, especially through 'The Cornwall Plan 2020–50' (Cornwall Council 2024). The report 'Sustainable Development in Cornwall; Local Perspectives on challenges and Opportunities' (Turner et al. 2021) discusses the application of doughnut economics for assessing local-level sustainable development, acknowledging that studies for local initiatives for sustainable development are established in research on the circular economy and transitions, and the use of the doughnut economics frameworks sits within this tradition. Importantly, the report identified that '…policy development at the county scale overlays a great deal of diversity at the local levels' (Turner et al. 2021). It is this diversity that is addressed in this book, and whilst the research in this book predates the conception of doughnut economics, many of the themes identified within the doughnut framework are presented here. However, these issues are identified in a less prescriptive manner and are explicitly drawn into the broader theoretical frame of a risk society, the importance of doing this has already been discussed. Moreover, by doing so, the research presented here overcomes some of the criticism of local applications of the doughnut economics framework. These include the doughnut not being able to capture complexity and connectivity across scales. Also, whilst the doughnut does appear to challenge the dominant economic growth paradigm, the prioritising of scientific expertise and policy focus of the doughnut reduces the ability to connect across spatial scales '… that has been achieved by the UN in the case of the SDGs' (Turner & Wills 2022:4). And extending this analysis it is argued that the doughnut is limited in its ability to navigate power dynamics, inequalities, and trade-offs.

Consistent with the overarching reflexive modern analysis of this book which emphasises uncertainty and contested knowledges formations, Turner and Wills suggest, in a post-functionalist paradigm, that it is the engagement of people's feelings and interpretations of their communities and their connection to place that underpins the ability of local communities to respond to the challenges of sustainable development. This aspiration is achieved in the research presented in subsequent chapters, with an emphasis on understanding sustainable development as a lived experience from individuals and communities. Against this backdrop the following discussion explores the relevant policy, legislative, and regulatory context that can restrict or enable sustainable development engagement for local government.

Local Government Context

In the following discussion we will outline the relevant legislative and policy frameworks as they relate to parish councils' capacity to embed sustainable development in their communities. This includes the Localism Act 2011, Environment Act 2021, and Climate Change Act 2008. There is also a focus on climate action and local energy governance, where there is an examination of the role of community-led renewable energy initiatives, decarbonisation policies, and local adaptation strategies. The following section focuses on community engagement and participatory governance, analysing the mechanisms of public participation, governance inequalities, and challenges in sustaining local engagement. This is then followed by looking at economic and social sustainability in local development, assessing local economic resilience, inclusive growth, and socio-economic disparities in sustainability implementation. The final section of this chapter identifies barriers and systemic contradictions, critically evaluating the multilevel governance conflicts, financial constraints, and contradictions between growth and sustainability objectives.

Whilst stated already it is worth repeating that town and parish councils are key but often overlooked actors in local sustainable development governance. Their proximity to communities, engagement in land management, and involvement in local planning position them as critical sustainability agents. However, they operate within a system of structural constraints, including funding limitations, governance ambiguities, and policy contradictions. By adopting a systems approach, this chapter aims to provide a connected understanding of how town and parish councils navigate sustainability governance, and whilst this serves to contextualise the rest of this book, this chapter also provides a stand-alone analysis that will provide critical insights for policy-makers, scholars, and practitioners.

The Structural Context of Local Sustainable Development Policy

Within the context of sub-political dynamics, multilevel governance and deliberative/participatory politics previously discussed, the legal and policy environment within which town and parish councils operate is shaped by national, regional, and international frameworks that impact the extent to which local councils can engage with sustainable development. Whilst these local councils have a degree of autonomy, their powers remain largely discretionary rather than statutory, meaning they often operate within a delegated responsibility model rather than one of genuine local self-governance (Pill 2022; Clarke & Cochrane 2013). As such, the following discussion critically examines some key legislative and policy frameworks affecting local sustainability governance.

The Sustainable Communities Act 2007

The Sustainable Communities Act 2007 sought to empower local communities through local authorities prioritising measures to enhance economic, social, and environmental well-being. The Act requires national government to consider local proposals in order to determine priorities and address specific issues. The Act recognised that sustainable development should be rooted in local contexts and shaped by local voices. The Act has been positioned within the context of devolution, strengthening local governance structures (Featherstone et al. 2012), making sure that community actors are active agents in delivering aspects of sustainable development. Within the Act sustainable communities are located within the broad categories of local, economic, environmental protection, social inclusion, and democratic involvement. These categories are attempts to operationalise sustainable communities through local participation. Clearly the ambassadors on locally defined priorities present the tensions between international and national policy discourses of sustainable development and the lived realities of communities. We explore this in more detail in Chapter 8.

The Localism Act 2011

This legislation was intended to devolve power to local communities, introducing mechanisms, such as neighbourhood development plans (NDPs), which allow local councils to influence planning decisions. The general power of competence (GPC) grants councils more flexibility in policy initiatives. Furthermore, the community right to bid enabled communities to protect local assets such as parks, pubs, and heritage sites. Whilst the rhetoric of localism emphasised community empowerment, in practice, the Act has been seen to be deeply contradictory, functioning as a mechanism for offloading responsibilities without sufficient financial and legal support (Pike et al. 2016).

Neighbourhood Development Plans

The introduction of NDPs was intended to enhance local participation in spatial planning, yet their effectiveness varies significantly. It has been identified that wealthier parishes with higher levels of civic engagement have successfully used NDPs to shape sustainable local development (Parker et al. 2015), but on the opposite end of the spectrum, more deprived areas, lacking technical expertise and financial resources, struggle to create and implement NDPs (Bradley 2018, 2020). A key limitation relates to the advisory rather than legally binding nature of the plans, meaning that developers and higher-tier planning authorities can override them if they conflict with broader economic growth objectives (Tait & Inch 2016). This represents a policy contradiction where the Localism Act promotes bottom-up governance, yet local plans are often subordinate to national growth targets, creating governance tensions between local sustainability priorities and centralised economic objectives (Clarke & Cochrane 2013).

The Environment Act 2021

The establishment of the Environment Act 2021 has introduced legally binding obligations for localities in a number of areas. This includes biodiversity targets, air quality standards, and waste reduction strategies, and the legislation has a direct impact on the way that town and parish councils engage with sustainable development-related issues. This includes biodiversity net gain, where new developments must evidence a 10 per cent net gain. There is an expectation that ecological enhancements move beyond existing conditions (DEFRA 2023; Ronish & Hilburn 2022). Also, on a multilevel governance basis, local nature recovery strategies ask councils to contribute to regional nature recovery networks. And Environmental Impact Assessments (EIAs) which strengthen regulations requiring a more rigorous environmental assessment for planning applications. Whilst biodiversity in land-use planning may be welcome, especially in regions such as Cornwall with high ecological value, they face significant enforcement challenges, as they lack legal mechanisms to enforce biodiversity net gain targets and combined with this are reliant on developers' voluntary compliance. This can ultimately lead to concerns about greenwashing and tokenistic sustainability claims.

So whilst the Environment Act strengthens the role of local authorities in conservation, it does not provide additional funding or administrative resources. This creates a policy-practice gap, where councils are expected to monitor and enforce environmental standards, but without dedicated staff or expertise. And this scenario is compounded by conflicting priorities between economic development and conservation that result in compromised sustainability outcomes (Raymond et al. 2023). With these observations in mind, it is sensible to suggest that whilst the Act strengthens environmental standards on paper, it struggles to address the structural barriers to local enforcement, which, as we have discussed, inevitably leads to uneven implementation (Davies et al. 2021).

The Climate Change Act 2008

The Climate Change Act 2008 has established legally binding carbon reduction targets towards net-zero emissions by 2050. Whilst there is a focus at the national level on large-scale energy transitions, at the local level, there is an expectation to embed climate adaptation strategies. This is not least as town and parish councils are well-positioned to implement community-based adaptation measures that may include flood resilience projects, particularly in coastal regions, community energy schemes, such as local solar and wind farms, and sustainable transport initiatives, including active travel networks. However, unsurprisingly decarbonisation policy remains highly centralised, with limited local control over energy governance. Moreover, community-led energy projects face regulatory and financial barriers, preventing their large-scale adoption (Creamer et al. 2019a, 2019b). Ultimately, national climate policies set ambitious targets but fail to provide town and parish councils with the necessary authority and resources to implement local energy transitions (Devine-Wright 2019).

The National Planning Policy Framework

The National Planning Policy Framework (NPPF) outlines national planning priorities, with a presumption in favour of sustainable development. However, critics suggest that the framework prioritises economic growth over environmental protection, leading to a number of negative and contradictory outcomes. This can include, for example, development approvals that override local sustainability concerns, which, combined with weak enforcement of environmental standards, particularly for new housing projects, contribute to negative environmental impacts. Policy tension between development and sustainability goals has been identified as contributing to the loss of greenbelt land to speculative development and erosion of local democratic planning processes, as national priorities take precedence (See Kirby & Scott 2023). This situation can be compounded by a lack of accountability for sustainability targets in the face of regulatory inconsistency. The NPPF frames sustainable development as a guiding principle, but, in practice, it can be argued, shifts decision-making power towards large-scale economic interests (Parker & Salter 2017).

Comparative Perspectives on Local Sustainability Governance

International comparisons highlight alternative governance models that provide greater legal and financial autonomy to local councils. Germany's decentralised energy policy enables local municipalities to own and operate renewable energy infrastructure. Scandinavian local governance models integrate sustainability into regional planning with dedicated climate adaptation funds. The Netherlands' water governance framework provides local councils with legal authority to enforce flood resilience measures. These models demonstrate how structural reforms in England could empower town and parish councils to take a stronger role in sustainability governance.

It is reasonable to acknowledge that the legislative and policy landscape surrounding town and parish councils is marked by contradictions and governance tensions. Whilst councils are increasingly expected to lead on sustainability, they lack statutory enforcement powers, financial autonomy, and regulatory support. Addressing these contradictions fundamentally requires legal empowerment to strengthen the enforceability of local sustainability policies, as well as funding to enhance outputs. The following section extends this analysis and moves to explore climate action and related energy governance at the local government level. As will be explored in subsequent chapters, issues relating to climate change and narratives of global warming explore multiple tensions in these areas.

The Role of Local Councils in Climate Governance

There is little doubt that climate change requires multi-scalar governance responses, tentatively balancing national and international policy goals with localised adaptation and mitigation strategies (Bulkeley & Newell 2010; Ostrom 2010). The conservative government has emphasised the need for the world to be climate-resilient by 2030. Since the Labour Party took power, new policy initiatives are slow to emerge. However, policy shifts at the national scale are evident in relation to housing and the NPPF (MHCLG 2025), which will have an impact on future climate resilience. Other areas of policy that are currently being explored include clean energy targets. The government has updated targets to achieve at least 95 per cent clean energy by 2030. The government has also launched Great British Energy (GBE 2025) with the goal of generating 8 GW of renewable energy by 2030. The Climate and Nature Bill was introduced in October 2024, and whilst not passing into law, the bill received cross-party support and civil society engagement. The proposed, and somewhat elusive, Green Prosperity Plan includes investments into green hydrogen, carbon capture, green steel, and creating 650000 new jobs by 2030 through climate and energy infrastructure (Financial Times 2025). And whilst the UK government has set legally binding net-zero targets under the Climate Change Act 2008, the mechanisms for delivering climate action at the local level remain fragmented and underfunded (Pike et al. 2016).

With the aforementioned in mind, town and parish councils can play an important role in climate governance in a number of ways. This could include community-led renewable energy projects (solar, wind, hydro), climate adaptation initiatives, particularly in coastal and flood-prone areas, sustainable transport and mobility schemes, and grassroots climate awareness and behavioural change campaigns. Despite these opportunities, and as consistent themes that can be applied to multiple areas of town and parish councils' operations, structural constraints, such as regulatory barriers, financial constraints, and lack of direct decision-making power, can serve to limit the extent to which local councils can lead meaningful responses to climate risks (Creamer et al. 2019a, 2019b).

With these persistent constraints in mind, the following discussion will explore some of the contradictions that exist between the national climate policy scope and local-level sustainability governance. Issues explored will include the structural challenges of local energy governance, particularly grid constraints and financial barriers. The role of community-led energy projects and the potential barriers they face will also be introduced. In light of the structural issues identified, the following discussion aims to explore the capacity of town and parish councils to lead climate adaptation and resilience-building efforts. These observations lead to conclusions relating to institutional reforms to empower local councils in climate governance.

The UK's Centralised Energy System and Local Disempowerment

The UK's energy governance model is highly centralised, with key decisions about energy generation, pricing – so – about energy generation, pricing , and grid infrastructure made at the national level. The national grid, for example, is not set up for significant decentralised energy generation, making it difficult for local projects to integrate with national supply (Blondeel et al. 2021; also see Roberts et al. 2014). Historically, privatisation of the UK's energy sector through the 1980s and 1990s established a foundation for regulatory complexity, especially in relation to planning regulations that actually favour large-scale developers instead of community energy groups (Creamer et al. 2019a, 2019b). And ever-present are the financial constraints for initial investment and capital costs relating to energy projects, as well as relevant and accessible funding mechanisms. With that said local renewable infrastructure has developed with a degree of independence from national energy policy, but still remains constrained by centralised regulation and pricing structures. As a result of this local councils have limited agency in ensuring that where local generation does exist, the outputs benefit local communities rather than larger corporate providers (Bridge et al. 2018; Seyfang et al. 2013; Hargreaves et al. 2013).

Taken together, these observations indicate that despite local councils' significant potential to facilitate the transition to low-carbon forms of energy, community-run projects face significant systemic barriers (Mirzania et al. 2019). This evolution of the energy landscape in the UK is not typical from a European perspective. For example, Germany and Denmark have allowed local councils to own and operate their own renewable energy infrastructure. The result of this decentralised approach has been more localised control of investment and energy pricing as well as greater resilience against energy market fluctuations (Moss et al. 2015). Ultimately, the UK's centralised and top-down approach to energy governance poses significant barriers for localised community-driven transitions.

Community-Led Renewable Energy Projects

With the context set, we can now move to the role of town and parish councils in the process of local energy generation more directly. It has been recognised that local councils have played a pivotal role in supporting and advocating for community-led renewable energy projects, particularly in rural areas with strong social cohesion (Hargreaves et al. 2013). Drawing on strategic niche management theory, Hargreaves et al. (2013) identify the role intermediaries play in catalysing grassroots initiatives and the coordination of place-based dynamics. As the authors point out, local authorities '… also often acted as intermediaries, supporting groups with planning and regulatory processes' (2013:878). Yet despite the opportunities energy projects in England remain small-scale and underdeveloped, with a lack of long-term policy certainty (Devine-Wright 2019). As we have already seen, difficulties accessing the national grid, loss of government support schemes, such as the Feed-in Tariff, which previously incentivised local renewable energy production, lead to contradictions in national policies that hinder local government action.

So, the persistent tension remains at a systemic level. Whilst national policies and narratives promote decarbonisation and increased energy resilience in the face of geopolitical and socio-economic shocks, the financial and regulatory system is skewed towards corporate-led projects that eclipse community-led solutions. It is within this context that town and parish councils must operate, and it is within this context that local councils are forced to consider their position for enacting solutions in relation to sustainable development.

Place-Based Climate Adaptation Strategies

As the International Panel on Climate Change recognises, climate adaptation, particularly in flood-prone, coastal, and rural communities, necessitates locally tailored solutions (Intergovernmental Panel on Climate Change 2022. And despite the structural and systemic issues we have already discussed town and parish councils are in a position to play a vital role in facilitating responses and solutions to climate change. For example, local councils are centrally placed to assist with the development of flood defence strategies, particularly in low-lying areas. Local councils can help facilitate strategies for protecting biodiversity and enhance and value ecosystem services as part of a drive towards nature-based solutions through collaborative governance processes (Garvey & Paavola 2022; Raymond et al. 2017). Local councils can also engage communities in disaster preparedness and resilience-building efforts (Raymond et al. 2023).

The use of nature-based solutions is increasingly recognised as being cost-effective and ecologically sustainable approaches to climate adaptation, for example, through flood management plans. Examples of local councils engaging in nature-based solutions can include wetland restoration to mitigate flood risk, reforestation and afforestation projects to enhance carbon sequestration and green infrastructure planning to create urban cooling effects. Cornwall Council has a dedicated nature-based solution programme that sits within its Climate Emergency Action Plan (Cornwall Council 2023a). The plan focuses on rewilding and biodiversity corridors, tree planting and woodland expansion, salt marsh and wetland restoration, and peatland restoration. Moreover, Cornwall Council is a pilot area for the Local Nature Recovery Strategy, which is supported by the Department for Environment, Food & Rural Affairs (DEFRA). The strategy promotes collaboration between Local Nature Partnership, Town and Parish Councils, and Landowners and Communities (DEFRA 2023). Other initiatives include Forests for Cornwall, which promotes engagement with town and parish councils (Cornwall Council 2023b). Further projects engage town and parish councils directly, for example, Pollinator Action Plans, St Just 2025. Whilst national climate adaptation policies acknowledge the need for local action, there is a lack of funding at the local level particularly for town and parish councils with direct funding or decision-making power to implement large-scale resilience strategies.

Strengthening Local Climate Governance

So whilst town and parish councils are strategically positioned to play an important role in facilitating and encouraging energy transitions, their ability to do so is limited by multiple structural barriers. This includes the centralised structure of UK energy governance, regulatory and financial obstacles that prevent the scaling of community energy initiatives. The lack of dedicated funding poses a significant barrier despite the increasing need for localised implementation. Without these issues being addressed, local sustainability responses will struggle to move forward in coherent and impactful ways.

The Role of Community Engagement in Sustainable Development

Whilst the technical, regulatory, and financial context is essential for understanding the opportunities and barriers for town and parish councils to engage with sustainable development, understanding the social and political contexts of sustainable development is fundamental for achieving a sustainable development. At the outset of this chapter we contextualised these areas within a sub-political, participatory governance framework. Concisely, the effectiveness of town and parish councils in delivering sustainable development-related initiatives is significantly impacted at the level of community participation, social cohesion, and local capacity for collective action (Fung 2006; Aylett 2013). Community engagement is essential in decentralised governance models, where local councils must mobilise public support, coordinate voluntary action, and navigate political tensions. However, it is fundamentally recognised that participatory governance is often unequal and contested, shaped by historical inequalities, socio-economic disparities, and differing levels of civic engagement across communities (Fung 2015; Bramley & Fitzpatrick 2018).

Local Intervention and Housing

Considering these observations we are able to identify multiple ways that town and parish councils are able to contribute to local sustainable development. For example, supporting local businesses through procurement policies, economic incentives, and business forums; promoting sustainable tourism, particularly in areas with environmental and cultural heritage value; facilitating community wealth-building strategies; ensuring local economies retain value rather than extracting it for external stakeholders. But the complexity of local intervention can also present barriers to the effective implementation of policies intended to address aspects of sustainable development. The final section of this chapter therefore offers a brief case study in this regard with a focus on housing, which, as we shall see in subsequent chapters, emerged as a key issue within the research.

In rural and coastal areas the issue of housing and affordability is ever-present, and we will revisit this and many of the issues raised in the discussion so far. Housing is central to economic activity and community cohesion. Rising property prices, especially in the sought-after coastal areas, continue to rise, putting home ownership out of reach for many people. This has led to local displacement and the rise in second-home ownership, as well as out-of-reach rental prices. This is compounded even further by the dominant natures of economic activities in particular areas, with a reliance on seasonal tourism and a transitory labour force. And whilst local councils establish initiatives that are designed to preserve and enhance the economic health of local communities and protect local affordability, it is recognised that this can lead to unintended econoHowever, mic consequences (Bradley & Sparling 2017).

Examples of this include the second home ban in St Ives. Through a referendum the community established a policy that prohibited the sale of new build properties for second home purposes. The clear intention was to prioritise housing needs for local residents and maintain the integrity of the local community. However, the impact of the policy had had unintended consequences. Prices for new homes in St Ives declined, and there was a shift in demand for existing properties.

This ultimately meant that affordability of the existing housing stock was increased further. The declining value of new build properties resulted in a lack of investment in new stock, reducing construction and impacting employment in the area. Concerns were also raised about the impact this was having on tourism in the area (The Economist 2019; Inside Housing 2017). And whilst each community has its own unique socio-economic and geographical distinctiveness, the example of St Ives serves to illustrate the complex and unintended consequences that can occur through even the most well-intentioned of policy initiatives. Looking at this issue in depth, Gallent et al. (2019) draw the following conclusion. Local governments must deal with the consequences of upstream decisions where regulatory frameworks are created by supranational and national bodies. In this instance, the established practice and economic consequence of second home ownership. But the authors acknowledge that local government must be seen to act on the pressures at the local level.

And this contradiction between local imperatives and broader national and supranational structural barriers extends beyond housing to impact multiple areas of policy-making. As we have already identified in Chapter 2, adopting a sustainable development perspective requires exploring the interconnections between the social, environmental, and economic dimensions of local and global dynamics which necessitates a move away from framing the economic resilience of an area through a growth-orientated lens. For local government these dynamics manifest in tensions between attracting investment and maintaining environmental integrity, supporting employment whilst reducing carbon-intensive industries, and encouraging tourism without overburdening local ecosystems.

And with that in mind there is evidence of a shift towards a more integrative approach that has moved from broader conceptual and theoretical discussions into policy frameworks. The rise of frameworks based around a circular economy that aims to minimise waste in the economic system, the use of ecosystem service valuation methodologies to incorporate the value of the services that we gain from nature. Multiple terms emerge that suggest an expanded and at the same time more integrative approach to economic progress. Ideas relating to a foundational economy which refers to local goods and services. This could include food and agriculture, renewable energy and water management, health, education, and transport infrastructure. And more broadly we can relate these discussions to economic models that sit both alongside existing neoliberal economic structures and models that sit and argue for the need to sit outside of these systems.

So whilst the previous discussion and analysis have taken a whistle-stop tour through relevant environmental, economic, and social areas that contextualise the operating space of town and parish councils, we must return to the integrated systems and sustainable development perspective that recognises the interconnectivity between economic, environmental, and social policy. Town and parish councils are central to economic sustainability, yet they face the embedded contradictory pressures that exist within the sustainable development narrative. The drive for economic growth whilst maintaining environmental integrity supports businesses whilst ensuring social equity, enhancing economic resilience within a system of centralised control. Whilst it is the overall contention of this book that town and parish councils are ideally positioned as key actors in local sustainable development governance, their effectiveness is constrained by systemic contradictions, institutional barriers, and governance fragmentation that we have explored in this chapter.

Conclusion

This chapter has provided a context from which we can understand the current and historical relationship local councils have with the broader social, political, environmental, economic, and legislative framework. This chapter has critically explored both the potential and the structural constants for town and parish councils, operating as catalysts for change towards a sustainable development. And despite some of the significant barriers we have already discussed these local government organisations are ideally placed to enhance sustainable development in their local areas. Reimaging sustainable development governance requires not only policy reforms but a much deeper shift in how we conceptualise power, economy, and democracy at the local level. Town and parish councils, if properly empowered, have the potential to become key actors for sustainable development. In the following chapter we explore that statement in more detail.

Bibliography

Aylett, A. (2013). Networked urban climate governance: Theoretical perspectives. *Local Environment*, *18*(4), 409–423.

Blondeel, M., Bradshaw, M. J., Bridge, G., & Kuzemko, C. (2021). The geopolitics of energy system transformation: A review. *Geography Compass*, 15(7), 1–22.

Borne, G. J. (2006) Sustainable development: the reflexive governance of risk (Thesis, University of Plymouth).

Bradley, Q. (2018). Neighbourhood planning and the production of spatial knowledge. *Town Planning Review*, *89*(1), 23–42.

Bradley, Q. (2020). The use of direct democracy to decide housing site allocations in English neighbourhoods. *Housing Studies*, *35*(2), 333–352.

Bradley, Q., & Sparling, W. (2017). The impact of neighbourhood planning and localism on house-building in England. *Housing, Theory and Society*, *34*(1), 106–118.

Brandsen, T., & Pestoff, V. (2006). Co-production, the third sector and the delivery of public services: An introduction. *Public Management Review*, *8*(4), 493–501.

Bramley, G., & Fitzpatrick, S. (2018). Homelessness in the UK: who is most at risk? *Housing Studies*, *33*(1), 96–116.

Bridge, G., Özkaynak, B., & Turhan, E. (2018). Energy infrastructure and the fate of the nation: Introduction to special issue. *Energy Research & Social Science*, *41*, 1–11.

Bua, A., & Bussu, S. (Eds.). (2023). *Reclaiming participatory governance: Social movements and the reinvention of democratic innovation*. Taylor & Francis.

Bulkeley, H., & Betsill, M. M. (2003). *Cities and climate change: Urban sustainability and global environmental governance*. Routledge.

Bulkeley, H., & Newell, P. (2010). *Governing climate change*. Routledge.

Bussu, S., Bua, A., Dean, R., & Smith, G. (2022). Introduction: Embedding participatory governance. *Critical Policy Studies*, *16*(2), 133–145.

Clarke, N., & Cochrane, A. (2013). Geographies and politics of localism: The localism of the United Kingdom's coalition government. *Political Geography*, *34*, 10–23.

Copus, C., Wall, R., & Roberts, M. (2017). *Local government in England: Centralisation, autonomy and control*. Palgrave Macmillan.

Cornwall Council. (2023a). Nature-based solutions. Cornwall Council. www.cornwall.gov.uk/environment/climate-emergency/nature-recovery-and-natural-environment/nature-based-solutions/.

Cornwall Council. (2023b). *Forest for Cornwall engagement hub*. Let's Talk Cornwall. https://letstalk.cornwall.gov.uk/hub-page/forest-for-cornwall

Cornwall Council. (2024). The Cornwall Plan 2020–2050. Cornwall and Isles of Scilly Leadership Board. https://www.cornwall.gov.uk/the-council-and-democracy/your-council/the-cornwall-plan-2020-2050

Creamer, E., Allen, S., & Haggett, C. (2019a). 'Incomers' leading 'community-led' sustainability initiatives: A contradiction in terms? *Environment and Planning C: Politics and Space*, *37*(5), 946–964.

Creamer, E., Aiken, G. T., Van Veelen, B., Walker, G., & Devine-Wright, P. (2019b). Community renewable energy: What does it do? Walker and Devine-Wright (2008) ten years on. *Energy Research & Social Science*, *57*, 101223.

Davies, C., Chen, W. Y., Sanesi, G., & Lafortezza, R. (2021). The European Union roadmap for implementing nature-based solutions: A review. *Environmental Science & Policy*, *121*, 49–67.

Department for Environment, Food & Rural Affairs (2023). Local nature recovery strategies: Policy overview. Department for Environment, Food & Rural Affairs. https://www.gov.uk/government/publications/local-nature-recovery-strategies/local-nature-recovery-strategies

Devine-Wright, P. (2019). Community versus local energy in a context of climate emergency. *Nature Energy*, *4*(11), 894–896.

Dryzek, J. S. (2002). *Deliberative democracy and beyond: Liberals, critics, contestations*. Oxford University Press.

Featherstone, D., Ince, A., Mackinnon, D., Strauss, K., & Cumbers, A. (2012). Progressive localism and the construction of political alternatives. *Transactions of the Institute of British Geographers*, *37*(2), 177–182.

Financial Times. (2025). Decarbonisation, downsized: What became of Labour's Green Prosperity Plan. https://www.ft.com/content/6c0f13ec-ec01-4778-b975-d582acb6083f

Fox, R., Korris, M., Hay, C., & Stoker, G. (2012). Audit of Political Engagement 9. The 2012 Report: Part Two The media and politics (Doctoral dissertation, Hansard Society).

Fung, A. (2006). Varieties of participation in complex governance. *Public Administration Review*, *66*, 66–75.

Fung, A. (2015). Putting the public back into governance: The challenges of citizen participation and its future. *Public Administration Review*, *75*(4), 513–522.

Gallent, N., Hamiduddin, I., Stirling, P., & Kelsey, J. (2019). Prioritising local housing needs through land-use planning in rural areas: Political theatre or amenity protection? Journal of Rural Studies, 66, 11–20.

Garvey, A., & Paavola, J. (2022). Community action on natural flood management and the governance of a catchment-based approach in the UK. *Environmental Policy and Governance*, *32*(1), 3–16.

Gelderblom, D. (2018). The limits to bridging social capital: Power, social context and the theory of Robert Putnam. *The Sociological Review*, *66*(6), 1309–1324.

Giovannini, A., & Griggs, S. (2024). Addressing the elephant in the room: What's the local in local government studies? *Local Government Studies*, *50*(6), 989–997.

Great British Energy. (2025). https://www.gbe.gov.uk/

Hargreaves, T., Hielscher, S., Seyfang, G., & Smith, A. (2013). Grassroots innovations in community energy: The role of intermediaries in niche development. *Global Environmental Change*, *23*(5), 868–880.

Inside Housing. (2017, May 18). The St Ives second homes experiment: One year on. Inside Housing. https://www.insidehousing.co.uk/insight/the-st-ives-second-homes-experiment-one-year-on-50354

Intergovernmental Panel on Climate Change. (2022). Climate change 2022: Impacts, adaptation and vulnerability. Contribution of Working Group II to the Sixth Assessment Report of the Intergovernmental Panel on Climate Change [H.-O. Pörtner et al. (Eds.)]. Cambridge University Press.

Jones, A. (2020). *The resurgence of parish council powers in England.* Springer Nature.

Jordan, A., & Huitema, D. (2023). Polycentric governance. In *Routledge Handbook of Environmental Policy* (pp. 55–67). Routledge.

Kirby, M. G., & Scott, A. J. (2023). Multifunctional Green Belts: A planning policy assessment of Green Belts wider functions in England. *Land Use Policy*, *132*, 106799.

Local Government Association. (2025). Local government association. https://www.local.gov.uk

Lowndes, V., & Sullivan, H. (2004). Like a horse and carriage or a fish on a bicycle: How well do local partnerships and public participation go together? *Local Government Studies*, *30*(1), 51–73.

Moss, T., Becker, S., & Naumann, M. (2015). Whose energy transition is it, anyway? Organisation and ownership of the Energiewende in villages, cities and regions. *Local Environment*, *20*(12), 1547–1563.

Ministry of Housing Communities and Local Government. (2025). Proposed reforms to the National Planning Policy Framework and other changes to the planning system. https://www.gov.uk/government/consultations/proposed-reforms-to-the-national-planning-policy-framework-and-other-changes-to-the-planning-system/proposed-reforms-to-the-national-planning-policy-framework-and-other-changes-to-the-planning-system

Mirzania, P., Ford, A., Andrews, D., Ofori, G., & Maidment, G. (2019). The impact of policy changes: The opportunities of Community Renewable Energy projects in the UK and the barriers they face. *Energy Policy*, *129*, 1282–1296.

Mythen, G. (2004). *Ulrich Beck: A critical introduction to the risk society*. Pluto Press.

Ofcom. (2023). Online nation 2023 report. Ofcom. https://www.ofcom.org.uk/siteassets/resources/documents/research-and-data/online-research/online-nation/2023/online-nation-2023-report.pdf

Ostrom, E. (2010). Beyond markets and states: Polycentric governance of complex economic systems. *American Economic Review*, *100*(3), 641–672.

Parker, G., Lynn, T., & Wargent, M. (2015). Sticking to the script? The co-production of neighbourhood planning in England. *Town Planning Review*, *86*(5), 519–536.

Parker, G., & Salter, K. (2016). Five years of neighbourhood planning – A review of take-up and distribution. *Town & Country Planning*, *85*(5), 181–188.

Parker, G., & Salter, K. (2017). Taking stock of neighbourhood planning in England 2011–2016. *Planning Practice & Research*, *32*(4), 478–490.

Parker, G., Wargent, M., Salter, K., & Yuille, A. (2023). Neighbourhood planning in England: A decade of institutional learning. *Progress in Planning*, *174*, 100749.

Pike, A., Rodríguez-Pose, A., & Tomaney, J. (2016). *Local and regional development*. Routledge.

Pill, M. (2022). Neighbourhood collaboration in co-production: State-resourced responsiveness or state-retrenched responsibilisation? *Policy Studies*, *43*(5), 984–1000.
Putnam, R. D. (2000). *Bowling alone: The collapse and revival of American community.* Simon Schuster.
Raymond, C. M., Frantzeskaki, N., Kabisch, N., Berry, P., Breil, M., Nita, M. R., … & Calfapietra, C. (2017). A framework for assessing and implementing the co-benefits of nature-based solutions in urban areas. *Environmental Science & Policy*, *77*, 15–24.
Raymond, C. M., Hirsch, P., Norton, B., Scott, A., & Reed, M. S. (2023). Rethinking appropriateness of actions in environmental decisions: Connecting interest and identity negotiation with plural valuation. *Environmental Values*, *32*(6), 739–764.
Roberts, J., Bodman, F., & Rybski, R. (2014). *Community power: Model legal frameworks for citizen-owned renewable energy*. ClientEarth.
Ronish, Y., & Hilburn, H. (2022). Biodiversity–gaining ground? *Environmental Law Review*, *24*(1), 3–9.
Seyfang, G., Park, J. J., & Smith, A. (2013). A thousand flowers blooming? An examination of community energy in the UK. *Energy Policy*, *61*, 977–989.
Sintomer, Y., Röcke, A., & Herzberg, C. (2016). *Participatory budgeting in Europe: Democracy and public governance*. Routledge.
Skerratt, S., & Steiner, A. (2013). Working with communities-of-place: Complexities of empowerment. *Local Economy*, *28*(3), 320–338.
Steiner, A., & Farmer, J. (2018). Engage, participate, empower: Modelling power transfer in disadvantaged rural communities. *Environment and Planning C: Politics and Space*, *36*(1), 118–138.
Tait, M., & Inch, A. (2016). Putting localism in place: Conservative images of the good community and the contradictions of planning reform in England. *Planning Practice & Research*, *31*(2), 174–194.
The Economist. (2019, July 18). St Ives's second-home crackdown has unintended consequences. The Economist. https://www.economist.com/britain/2019/07/18/st-ivess-second-home-crackdown-has-unintended-consequences
Turner, R. A., Blundell, A., Collins, C., Exeter, O. M., & Wills, J. (2021). *Sustainable development in Cornwall: Local perspectives on challenges and opportunities*. Environment and Sustainability Institute, University of Exeter.
Turner, R. A., & Wills, J. (2022). Downscaling doughnut economics for sustainability governance. *Current Opinion in Environmental Sustainability*, *56*, 101180. https://www.sciencedirect.com/science/article/pii/S187734352200032X
Watkins, H. (2021). *UK localism in transition and the politics of community*. Bloomsbury Publishing PLC.
Wills, J. (2016). *Locating localism: Statecraft, citizenship and democracy*. Policy Press.
Willett, J. (2021). *Affective assemblages and local economies*. Rowman & Littlefield.

4 Parish Context

Cornwall

I have already identified my connection to Cornwall through family heritage and geographical location. There is, to a certain degree, something intangible about what makes Cornwall Cornwall. In subsequent chapters we will explore that theme through issues relating to place and identity. We can say that Cornishness represents a distinct cultural identity, and as with any culture, this refers to language, traditions, and historical contexts of heritage. In 2014, under the Framework Convention for the Protection of National Minorities, the UK government recognised Cornish people as a minority (Council of Europe 2025; UK Government 2014). This means that Cornish people share the same status as the Welsh, Scots, and Irish. In recognition of this status, Cornwall Council has initiated over 30 Cornish distinctiveness projects which sit under the broader Cornish culture initiative.

Geographically Cornwall is located in the South West of England. It has a dispersed settlement pattern that is predominantly rural and coastal. Cornwall has long-standing and well-recognised socio-economic challenges with persistently low wages, with pockets of significant deprivation, especially in former mining communities and isolated rural areas (Cornwall Council 2025). Seasonal tourism, agriculture, and fishing have formed the main components of the Cornish economy. In 2015, Cornwall was the first area in the UK to be granted a rural devolution deal, providing the county with new powers and funding to address embedded challenges (DLUHC 2023). The Vision 2030 strategy produced by the Cornwall and Isles of Scilly (Local Enterprise Partnership 2022) emphasises areas such as clean economic growth, renewable energy, digital connectivity, and climate resilience. Cornwall Council's own strategy on climate emergency action outlines a

Table 4.1 Key Statistics Cornwall

Statistic	*Cornwall*
Population	570,300
Area	354,779 hectares
Governance structure	Unitary
Sub-County delivery	19 network areas (at the time of data collection)
Parish councils	213

DOI: 10.4324/9780429297199-4

commitment to become carbon neutral by 2030 (Cornwall Council 2025). Cornwall represents an intriguing case study for exploring sustainable development, local governance, and community resilience in a peripheral rural region of the UK. Table 4.1 presents key statistics at the time of data collection.

Local Government Restructuring

During the data collection period Cornwall Council was in the process of transitioning to a unitary authority. This provided invaluable opportunities to explore transitional dynamics and the engagement with the level of government that is closest to local communities. The move towards a unitary authority involved the dissolution of the six district councils and the establishment of a single unitary authority and 19 Community Network Areas (CNAs). These network areas were designed to provide a bridge between the unitary authorities and the local communities they served, supporting a consultative and democratic process. Each network area had a community network panel that included Cornwall councillors and representatives from town and parish councils. Eventually these panels' effectiveness was evaluated, and Cornwall Council initiated a review that resulted in the 'Let's Talk Cornwall' initiative, which laid the groundwork for the creation of 12 larger Community Area Partnerships (CAPs). Table 4.2 outlines the key differences between these areas.

Alongside the aforementioned, structurally reformed town and parish councils are embedded in a network of wider representative organisations that support and shape local councils' governance capacity. The National Association of Local Councils (NALC) represents over 10,000 local councils and 100,000 councillors (NALC 2025). The organisation provides resources, training, and helps standardise policies and practices across diverse councils. Also NALC represents the sector at national government level, helping secure recognition for local councils within local policies and agendas. Complementing this representational role, the Society of Local Council Clerks (SLCC) provides resources for over 5,000 clerks and senior officers who administer town and parish councils (SLCC 2024). What is recognised here is that these broader organisations highlight that town and parish councils do not operate in isolation. Instead, they form part of a broader network of local governance in England. Their organisations can act as facilitating organisations that connect grassroots democracy with national frameworks and policy agendas.

Table 4.2 Area Comparison

Aspect	*Community Network Areas (CNAs)*	*Community Area Partnerships (CAPs)*
Number	19	12
Structure	Community network panels	Community area partnerships
Focus	Granular, local governance	Strategic, service-aligned regions
Establishment	2009	2022 onwards
Engagement	Parish/town councils and Cornwall councillors	Broader multi-agency collaboration

The Survey

The survey was dispatched to over 2,000 councillors across Cornwall; this represents all Cornwall parish councils, numbering approximately 213. Response rates were high, with between 50 per cent and 70 per cent of parishes responding to the survey. This is a very significant response rate, particularly in light of the length and complexity of the questionnaire (see Appendix 1). Initial introductory letters were sent out to all parishes explaining the purpose of the survey, as well as outlining the key areas to be covered. The survey was delivered by post, and respondents were asked to return the questionnaires collectively in a single prepaid envelope that was held by the parish clerk. Each questionnaire was numbered individually, and a record was kept of where each individual questionnaire was sent. Numbering individual questionnaires proved to be the most effective method of delivery. Postal information was not always available or correct, and early internet connectivity, especially in isolated rural areas. A small percentage of the questionnaires were returned as they had been sent to an incorrect address. On a number of occasions, subsequent investigations had to be made to identify the correct addressee. Also on a number of occasions a single clerk was responsible for more than one parish. Records of these situations were not always accurate or up to date. On these occasions clerks contacted the research team and requested further questionnaires which were sent as a matter of urgency.

Survey Design

As is already clear the overall aims of the research are multifaceted, and the questionnaire had to accommodate multiple issues, practically and theoretically. The basic format of the questionnaire design was drawn from earlier research that focused on the integration of sustainable development into governance frameworks at the local and individual levels (Borne 2006). The survey was structured to gain different levels of information from the respondents. Questions were designed to elicit basic answers on the survey group's background, but also more in-depth questions were formulated pertaining to environmental knowledge and awareness, issues of sustainable development, questions of responsibility, and personal agency. These issues were identified as being fundamental factors when assessing the relationship between individuals' attitudes and the relationship between knowledge of sustainable development and personal reflexive behaviour. Other studies have drawn on similar issues (Bamberg & Möser 2007; Otto & Pensini 2017), as well as broader topics in relation to communication, awareness, sustainability literacy, and the ways that these issues impact behavioural change and engagement with sustainable development (O'Neill & Nicholson-Cole 2009; Whitmarsh et al. 2011; Leal Filho et al. 2018).

On a wider scale, national and international surveys, including the UK Governments Public Attitudes Tracker (UK Government 2019) and the European Commission's Eurobarometer series on environmental issues (European Commission 2019), have demonstrated the value of survey research for mapping public knowledge and attitudes in areas of climate and sustainable development. These surveys have provided fertile ground for research in multiple areas. For example, it has been identified consistent gaps between awareness and action (Whitmarsh & O'Neill 2010). Moreover, Poortinga et al. (2019) have highlighted how concerns for climate change interact with social norms, trust, and perceptions of responsibility. The research explored in this book positions these themes at the micro-level of parish council governance where discourses of sustainable development are negotiated within the specific institutional and community contexts of Cornwall.

Whilst being informed by the areas already discussed, the additional dimension that needed to be integrated into the survey design for this research was to ensure that it was an effective vehicle not only for gathering relevant data on sustainable development-related issues, but also for being able to explore the relationship between sustainable development and a reflexive modernity. The early survey design was based on this premise and was successful at creating just such a vehicle. As we have already seen, these issues were identified as being fundamental factors when assessing the relationship between individuals' attitudes and the relationship between knowledge of sustainable development and personal reflexive behaviour. Relevant sections of the survey are included in Appendix 1. Table 4.3 distils the key themes in the survey and identifies the areas that relate to identifying a reflexive modernity.

These indicators form a framework for assessing the extent to which sustainable development represents a reflexive modernity within the data. This is not achieved in a prescriptive or quantifiable manner, but instead it is used as a guide for an overall qualitative assessment of the relationship between sustainable development and a reflexive modernity. Using the aforementioned structure the final stage for

Table 4.3 Survey and Reflexive Modern Indicators

Indicator of Reflexive Modernity	*Questionnaire*
1 Displacement of traditional authority	Reveals how much governance is devolved and how councils work with other institutions – signs of a shift from top-down to networked decision-making. Elements of parish planning, partnerships, and strategic goals.
2 Knowledge pluralism/ critique of expert systems	Highlights local views on whether local authorities understand or consider local needs. Identify resistance to imposed expert frameworks.
3 Ethical and moral reasoning	Explores whether sustainable development is framed in moral terms such as intergenerational equity or dismissed as jargon. Represents reflexive moral engagement or scepticism.
4 Sustainable consumption/reflexive choice-making	Captures personal and parish-level practices like recycling, awareness of carbon footprint, and behavioural change. Representative of reflexive environmental subjectivity.
5 Reflexive governance structures	Assesses whether councils use adaptive processes, for example, feedback from the community and work in evolving partnerships. Representative of experimental, reflexive governance.
6 Global-local risk awareness	Reveals whether global risks like climate change are seen as relevant to Cornwall linking local sustainable development actions to global reflexive concerns.

developing the survey, as presented here, was to tailor it specifically for town and parish councils. To that end, through desk-based research the survey was tailored to identify explicit issues that were of relevance to town and parish councils in Cornwall. The survey was also informed by an extensive consultation period with Cornwall Council and town and parish councils. Parish councils were also visited to understand what their priorities might be and how these related to local, national, and international imperatives. With the key areas of survey development explored the following discussion will discuss data analysis.

Survey Analysis

Survey data was input into the Statistical Package for Social Sciences (SPSS), and descriptive statistics were used to explore frequencies and compare themes and variables within the survey. It was not necessary for the purposes of the research to employ inferential statistics, however, there is significant scope to reassess the database to explore the relationship between variables. The survey also provided a vehicle to recruit participants for the interview component of the research. The subsequent interviews drew on and extended the survey format and underlying rationale, not in a structured way but in an in-depth and fluid assessment of the relationship between sustainable development and reflexive modernity. The following section will discuss the interviews in more detail.

Interviews

Interviews were conducted systematically throughout the 19 community network areas which were established by Cornwall Council as the council moved to unitary status in 2009 as part of a broader policy to connect local communities with decision-making process, following the removal of the six district councils. The aim of the interviews was to elicit more in-depth data than was possible in the questionnaire. This form of triangulation increases the validity and rigour of data collected. The questions posed in the interviews were drawn from and expanded upon the questionnaire format. As interviews were conducted the process evolved and responded to the realities of the interview process. Initially, the following protocols were observed. Once the interview was confirmed a copy of their questionnaire was sent to the interviewee to refresh their memories on the issues that were to be addressed in the project. An interview schedule was developed from the questionnaire, which was designed to respond specifically to the interviewee's survey responses on an individual basis. However, as the interviews progressed, it was observed that an overly structured approach stifled the interview process in a number of situations. It was decided therefore that a set of generic questions on the primary areas of the research would be used as a broad guide.

Analysis

Discourse analysis was applied to the transcripts. In this context discourse is understood as a specific '… ensemble of ideas, concepts and categorizations which are produced, transformed and reproduced in a set of social practices defining the meaning of the social and the objective world' (Hajer 1995:44). The resultant analysis occurs from a staticised text or transcript, transforming even the 'active' event of interviewing into a 'static text' (Holstein & Gubrium 1994). The following discussion will examine the nature of the transcript to identify the analytical processes used in the research presented here. Whilst the transcript is a static text, it is also a dynamic medium of analysis. For example, Kvale (1996) points to the hybridity of a transcript, suggesting that the transcript is neither a transparent record of speech nor neutral as a container of data. The transcripts' origins remain an active part of the subtext constructed by earlier methodological choices. In this sense transcripts actively mediate between the interview event, the text, and the subsequent analysis, raising critical questions of representation and reflexivity (Bucholtz 2000; Nikander 2008; Braun & Clarke 2021).

This hybridity means that it is essential that the researcher chooses the method of interpretation that is conducive to the research area. As Fairclough notes discourse analysis necessarily oscillates between a micro focus on specific texts and the wider context of discourse, that is, '...the relatively durable social structure of language which is itself, one element of a relatively durable structuring and networking of social practices' (2003:3). In this research, discursive representations of sustainable development and associated issues are examined not only as textual artefacts but also as products of the institutional and social environments from which they emerge. In so doing we are explicitly acknowledging the relationship between linguistic practice, socio-political context, as well as the reproduction and transformation of power relations (van DijK 2017; Wodak & Meyer 2015).

A Grounded Approach to Theory Development

The analytical tradition of a grounded approach to data analysis is to create an active and iterative process between empirical data and theoretical development (Corbin & Strauss 1990; Charmaz 2014; Bryant & Charmaz 2007). In this context the analytical process is ongoing, with each stage influencing the next. A grounded approach theory allows for versatility in the research process and allows data to be simultaneously examined both inductively and deductively, encouraging sensitivity to context and emergent categories (Clarke 2005; Thornberg 2012). At the centre of the research discussed in this book is a process of induction where discursive representations of sustainable development are derived from the data. However, these inductive findings are utilised to test and refine the central theoretical proposition, which is that discourses of sustainable development are representative of a reflexive modernity. So what we are doing is fundamentally recognising that the boundaries between induction and deduction do not represent an impenetrable barrier, and that researcher perspectives and pre-existing theoretical insights inevitably shape the construction of categories and themes (Charmaz 2020; Thornberg & Dunne 2019).

Explicitly acknowledging permeability between induction and deduction also suggests ways to strengthen the limitations of grounded theory. Silverman (1998) and Dey (1996) have argued that there is often a lack of acknowledgement of early theories and subjective input that guide the research process. In light of this what I emphasise here is the importance of recognising that there is a cyclical relationship between theory, life experience, the research procedure, analysis, and subsequent conclusions and recommendations. In this research reflectivity is especially heightened as the analytical focus directly addresses representations of sustainable development and the relationship to a reflexive modernity. With this in mind the interview format drew on the findings of the survey to gain a deeper understanding of key issues related to sustainable development, whilst also exploring reflexive modern indicators. And as with the survey design, the interview schedule was distilled into essential areas of enquiry. These included community, partnership, sustainable development, and key environmental issues, which have been identified as key issues for the exploration of the relationship between sustainable development and a reflexive modernity.

Within this context conversations were allowed to develop with interviewees expressing their views on a wide range of issues. Practically this meant that interviews were often lengthy in nature, typically ranging from one to two hours, though in some instances interviews lasted for up to four hours. Resulting transcripts ranged between 5,000 and 20,000 words. Data was analysed on a rolling and comparative basis, building up categories as the analysis evolved. During this process clear themes emerged on multiple issues. Some themes were expected, and others presented interesting insights into how sustainable development was being understood and embedded within a local community context. We will explore these themes in subsequent chapters. At this point I recognise that my own position as both analyst and participant in the discourse of sustainable development has inevitably shaped the research process and outcomes. I approached every stage of the research not as a linear and inevitable progression but as a cyclical and interactive process. From survey design to interview interaction, transcription, coding, and theorisation, it was all informed by my prior knowledge and subjectivities. In practice what this means is that the key categories identified in this book, community, partnership, and sustainable development, to name a few, were examined both as emergent categories from the data and as a result of my ongoing engagements with the concepts of both reflexive modernity and sustainable development. With the methodological approach discussed the following section will move to discuss the principal phases of the research, as well as introducing initial survey results.

Research Phases and Initial Results

A five-phase approach was adopted in order to elicit the necessary data. This included the initial consultation, the customising of the research design, initiation of the survey and interviews, as well as the overall analysis. These phases are presented in Table 4.4.

In the remainder of this chapter we begin to build a picture of the research participants and the research area. We will explore the demographics of the councillors that responded to the survey. This includes key social science variables such as gender, age, occupation, and qualifications. We also explore the respondents' role within their parish council. Following this we move to address some key understandings of the context from within which town and parish councils operate. The contextual issues include identification of settlement type, whether respondents felt that they identified as urban or rural. Also, initial insights into parish councils'

Table 4.4 Parish

Phase	*Focus*	*Key Activities*
Phase One: Initial consultation	Identification of research area	Contact with local authorities and organisations; recognition of town and parish councils as closest tier of government with community leadership role.
Phase Two: Adaptation	Tailoring design to Cornwall and global concerns	Adapted research design; consultation and focus groups with councils and stakeholders.
Phase Three: Survey	Data collection across Cornwall	Survey distributed to all town and parish councils; responses collated in SPSS database.
Phase Four: Interviews	In-depth exploration of issues	Interview schedule based on survey; one interview per council (95 total); all interviews recorded and transcribed.
Phase Five: Analysis	Data interpretation	Scrutiny of survey and interview data; themes developed through analytical process.

partnership arrangements. This includes both relationships with the county council and broader sector partnerships. The final part of this chapter will explore what parish councillors feel are the barriers to effective service provision which connects the role of partnerships to the operational duty of parish councils. As has already been discussed within the context of exploring the presence of reflexive modern processes, we address the displacement of traditional authority and knowledge pluralism through a critique of expert systems.

Parish Demographics

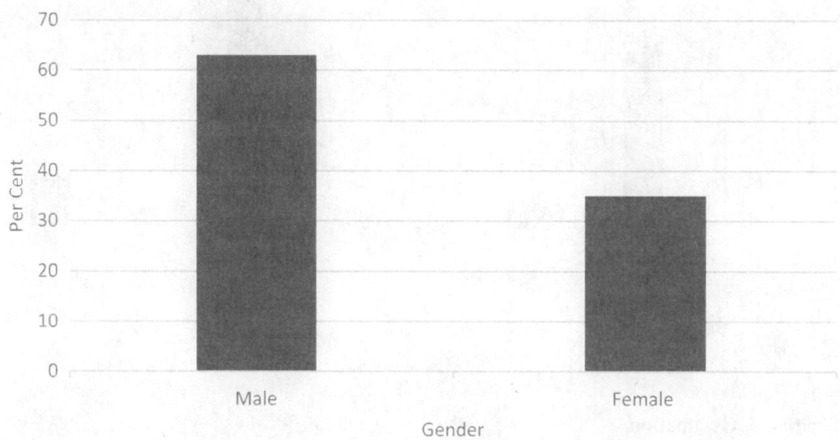

Figure 4.1 Gender.

Sixty-three per cent of respondents were male, and 35 per cent of respondents were female. It is recognised here that the narrow gender allocation could be expanded to include broader categories for future research (Figure 4.1).

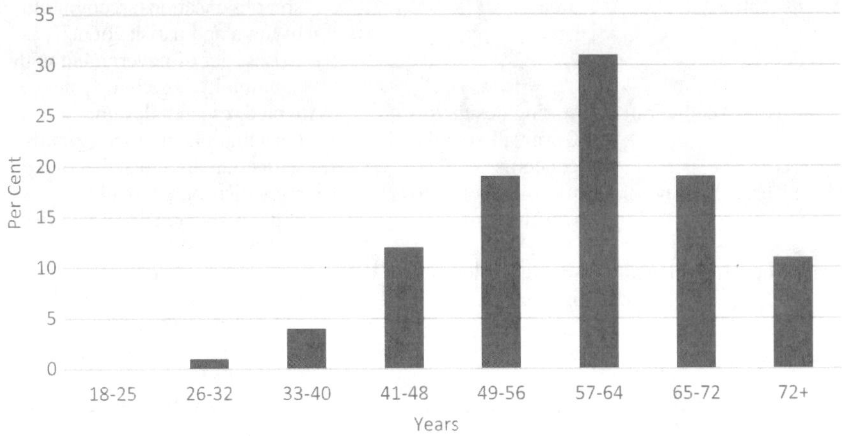

Figure 4.2 Age.

Respondents were asked to indicate their approximate age. As shown in Figure 4.2 only a very small percentage of respondents are below the age of 40. Over 40, 12 per cent of respondents were between 41 and 48; 19 per cent of respondents were between 49 and 56; 31 per cent of respondents were between 57 and 64; 19 per cent of respondents were between 65 and 72; and 11 per cent of respondents were 72+ (Figure 4.3).

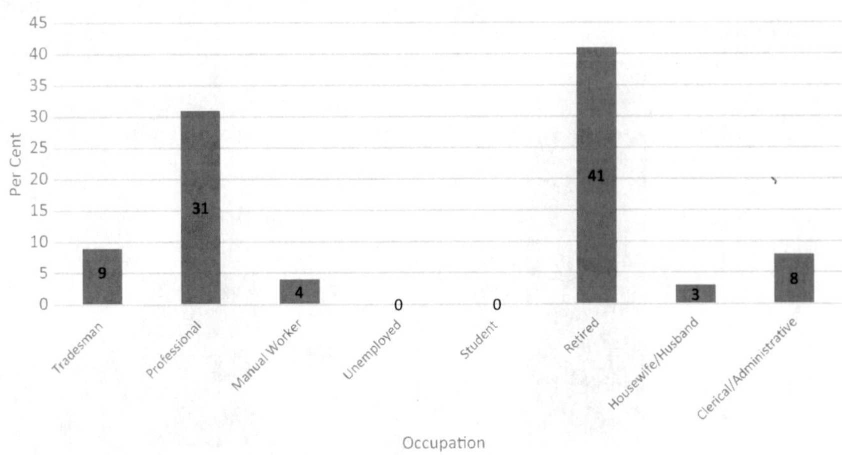

Figure 4.3 Occupation.

Members of town and parish councils were asked to identify their occupation. A total of 9 per cent said they were tradesmen; 31 per cent indicated that they were professionals; 4 per cent said that they were manual workers; no respondents were students; 41 per cent said that they were retired; 3 per cent indicated that they were housewives or househusbands; 8 per cent said that they were clerical/administrative, and none of the respondents were unemployed.

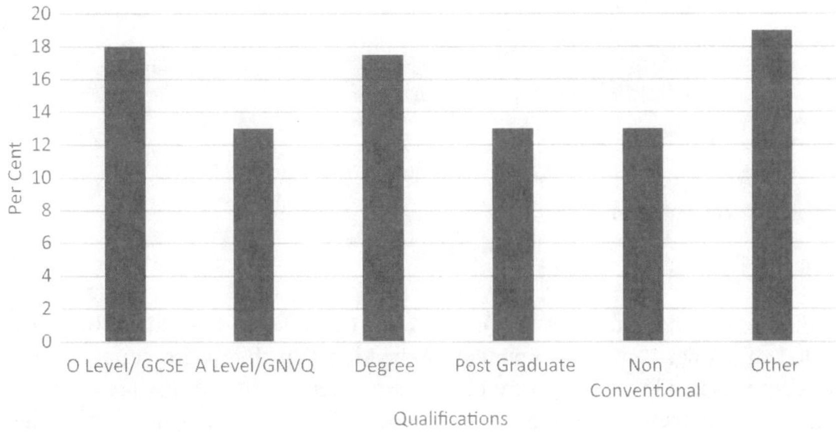

Figure 4.4 Qualifications.

Members of town and parish councils were asked to indicate their level of education. Eighteen per cent said that they held 'O' levels or General Certificates of Secondary Education; 13 per cent indicated that they held 'A' Levels or General National Vocational Qualifications; 17.5 per cent indicated that they had a degree; 13 per cent said that they had post-graduate qualifications; 13 per cent indicated that they had non-conventional qualifications; and 19 per cent said that their qualifications were other.

Taken together this demographic profile seems typical of wider trends in town and parish councils that are seen as having a skewed gender balance and an older professional membership. And whilst this wealth of experience can bring considerable expertise to local government it also highlights persistent issues relating to representation and diversity, particularly for gender and age. And whilst recognised that the gender imbalance has improved in recent years, these disparities still exist (Copus et al. 2017; Giovannini et al. 2023). Inevitably participation of parish councils is shaped by the structural and cultural barriers within which they sit. Barriers of participation can include time, political disengagement, and perceptions of what a parish council role entails, and it is within this context that we explore councillor narratives in the remainder of this chapter as well as subsequent chapters.

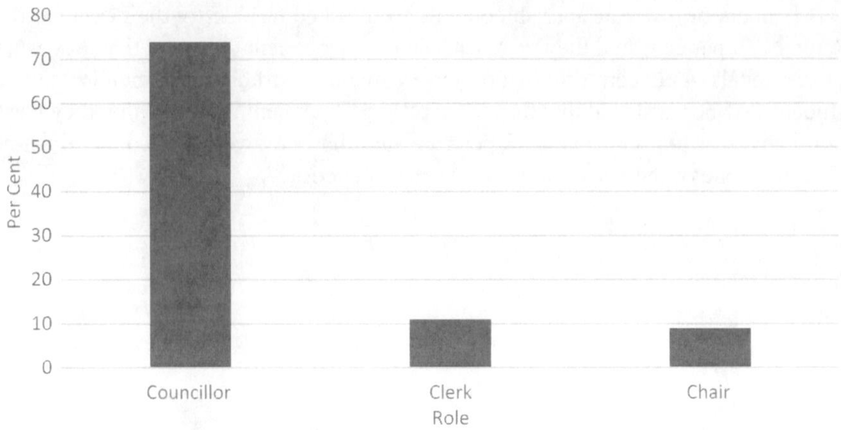

Figure 4.5 Role in Parish.

Figure 4.5 illustrates that the majority of respondents were councillors, with a total of 74 per cent of respondents indicating that they were councillors; 11 per cent indicated that they were clerks; 3 per cent indicated that they were the employees of their town and parish council. The interviews also represented this general spread, with the majority of interviewees being councillors. Subsequent analysis and data presented in the interviews are drawn entirely from town and parish councillors. This is primarily because councillors are the elected representatives serving on a voluntary basis. In this way councillors provide the primary democratic link between the council and the communities that they represent. By extension councillors' discourses and perspectives are central to understanding local governance and community identity. From a methodological point of view the data set overwhelmingly comprised of councillors, making them the most appropriate focus for analysis. With this in mind it is acknowledged that the clerks' role within a council is pivotal. Clerks provide professional advice on multiple issues, ensuring legal compliance and ensuring the day-to-day administrative elements of parish council activities are properly administered. Also, clerks provide an institutional memory and continuity within any particular council that can see multiple changes in councillor membership over a number of years (SLCC 2025). Further research would certainly benefit from the inclusion of clerks' perspectives.

Respondents were asked if they felt their council represented a community that was urban, rural, or mixed. The notion of rurality and the incumbent and persistent structural issues contained therein underpin a significant undercurrent of narratives throughout the research. Issues of poverty, marginalisation, isolation, and relocation are combined with narratives of natural beauty, freedom, heritage, and community (Shucksmith 2018). There is a continued need to provide an in-depth understanding of these constructs to avoid imposed and ineffective

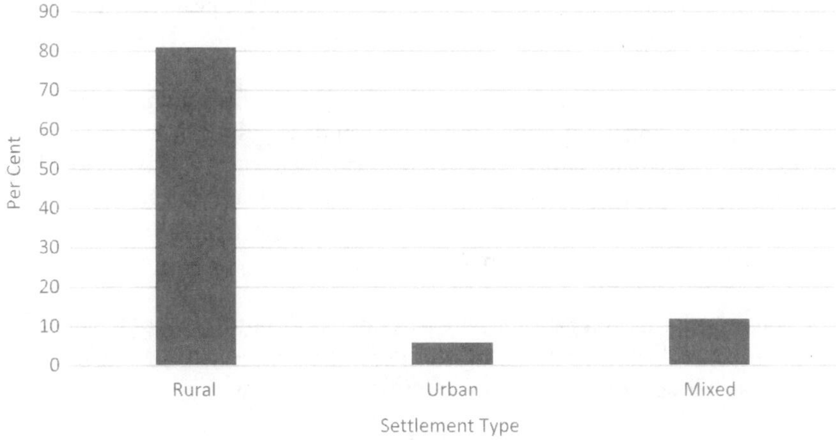

Figure 4.6 Settlement Type.

policies (Willett 2023a, 2023b; Woods 2010). The first step on that path is to identify how people and communities built identity and action around notions of urban and rural. As illustrated in Figure 4.6, the vast majority of town and parish council members felt their local area was rural. A total of 81 per cent said they lived in a rural area; 6 per cent said that they lived in an urban area; and 12 per cent indicated that their settlement type was mixed. Within the interviews there was a strong identity with rurality and often passionate proclamations about how the heritage and identity of Cornwall should be preserved. This is an important issue for understanding the nature of sustainable communities, as an increasingly urban global population fundamentally alters the rural landscape and the policy attention directed towards it. Whilst this is revisited in more detail in subsequent chapters the following interview data highlights the construction of rural and urban areas.

Whilst interviewees described the physical attributes of what constitutes their rural areas, rurality is a subjective construct and varies depending upon the perspective of the interviewee. Often where a councillor was part of a larger village or small town, there would be ambiguity over the nature of the settlement type. The following extract highlights this point as the interviewee is asked how the area being discussed is described:

> …well don't call it a town. They get very upset about that. It's a large large village and I personally would call it semi rural. The surrounding places I would say are probably more rural but actually if you look at how the development of (the parish) sort of spread a bit and I know there are boundaries that everything is now sort of, I mean I don't think there's much room for more development in (the parish). The infrastructure simply can't take it but other places in the parish could take a bit more. My personal feeling would

> be sort of semi rural. Its not like you are not looking at the peaks and the lakes but I think other people probably describe it as rural.

This extract exemplifies the ambiguous nature of settlement within Cornwall as the interviewee, whilst not using the term urban does describe the area as urban. The extract also indicates the strength of maintaining a 'rural identity' as the interviewee describes other councillors becoming 'very upset' about describing the area as a town. Indeed, even when interviewees from the larger conurbations were interviewed, their perspective on the areas remained rural:

> I would say it's a rural parish. its not a city parish by any stretch of the imagination any of them you know. The only difference, the only problem with this parish is it is so elongated so its very difficult you know

This interviewee reinforces the perception that even the larger parishes retain a rural identity. This is an identity which is intrinsic within the interview material and reflects the nature of community in Cornwall. There was also recognition of the changing nature of the landscape and the potential future urbanisation, as increased population pressures force authorities to consider housing priorities. For example, the following councillor was asked if he considered his parish urban or rural:

> Well we've always considered ourselves as rural but I think government now looks at us more as urban, if it goes the way they want it to go it will certainly be urban because there's plans to build thousands of houses down there.

And on a broader cross-cutting policy basis we are already seeing initial tensions between local and national government, imposed perceptions, and national policy imperatives. These tensions echo border debates where it is recognised that localised understanding of place clashes with national policy imperatives around housing and urban expansion (Allmendinger & Haughton 2010; Gallent 2014). These issues will be repeatedly revisited as we explore different elements of sustainable development throughout this book. With the background to the research area established and the demographics of the survey group presented, the following section moves to explore elements of the survey that highlight the way that town and parish councils interact with other stakeholders and organisations. By expanding on the way that councils understand partnerships we are able to initially explore the actual and aspirational governance structures within which councils sit.

Local Governance

Throughout this book so far, we have explored different dimensions of governance. It has been suggested that a central indicator of sustainable development's ability to reflect a reflexive modernity is the expansion of, and opening up of, governance processes at both the global and local scales. Working in partnership is a central facet of governance processes. In the local government context, partnerships exist both horizontally with neighbouring councils, as well as vertically with the

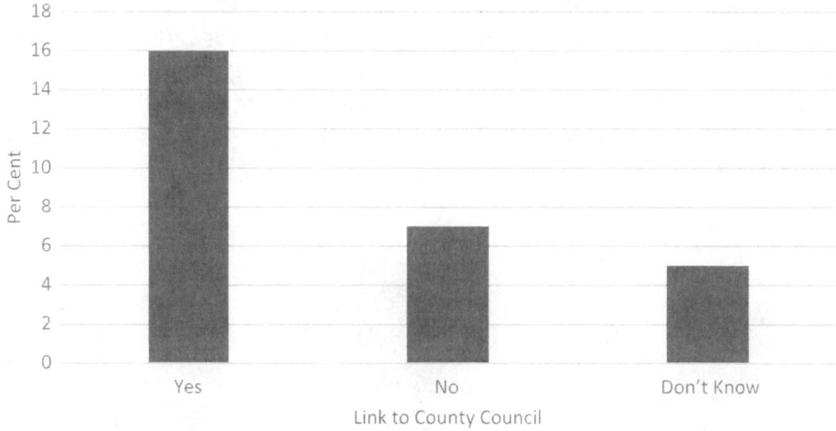

Figure 4.7 Linkages to County Council.

county council and other organisations across sectors. These partnerships are seen as essential in order to address complex needs and ensure the efficient delivery of shared services. Partnerships have the capacity to create mechanisms for councillors to pool resources, enhance local capacity, and influence and enhance strategic decision-making processes (Lowndes & Sullivan 2004; Cornwall Council 2023). The following section explores the elements of the survey that relate to partnership.

Respondents were asked if there should be a linkage to their county council. Figure 4.7 shows that 16 per cent indicated that there should be a link to the county council; 7 per cent indicated that there should not be a linkage between the parish and the principal authority. More significant than the indicated results however, and in contrast to response rates for other questions, 72 per cent of respondents chose not to respond to the initial question.

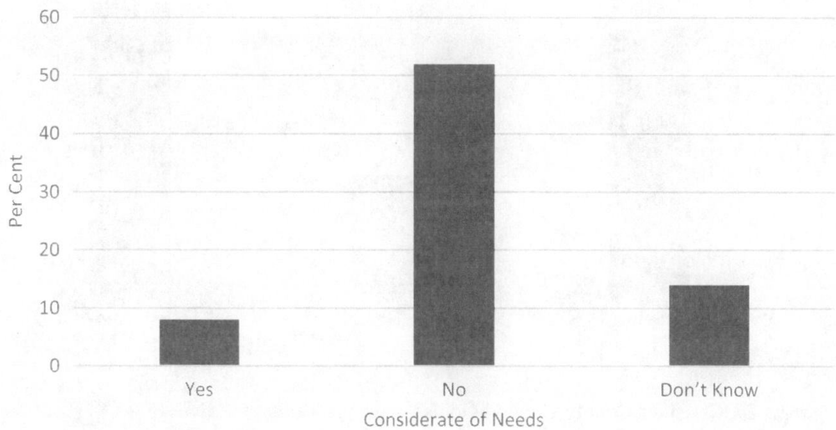

Figure 4.8 County Council Considerate of Local Needs.

Continuing with the relationship between the county council, respondents were asked whether they felt that the county council authority was considerate of local needs. Figure 4.8 shows that 52 per cent indicated that they did not feel that their principal authority was considerate of local needs; 8 per cent indicated that the principal authority did respond to local needs; and 14 per cent indicated that they did not know.

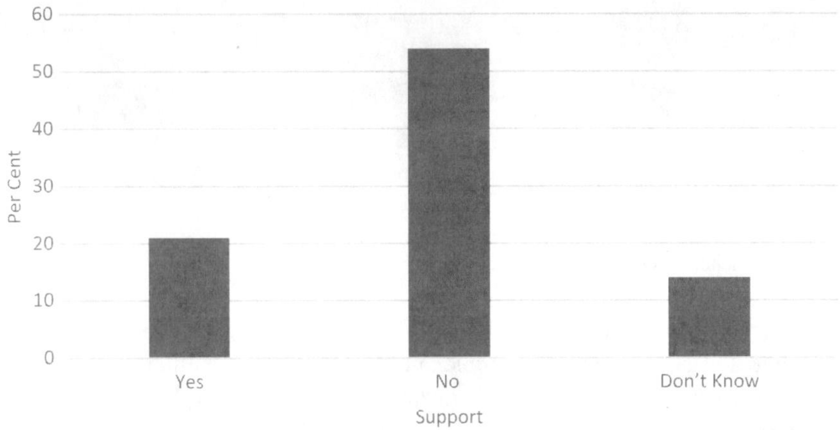

Figure 4.9 County Council Support Parish Council.

Similarly, respondents were asked to indicate if they felt that their county council was supportive of their parish councils. Figure 4.9 shows that the majority of respondents, at 54 per cent, indicated that the county council did not support their role as a parish councillor; 14 per cent indicated that they did not know.

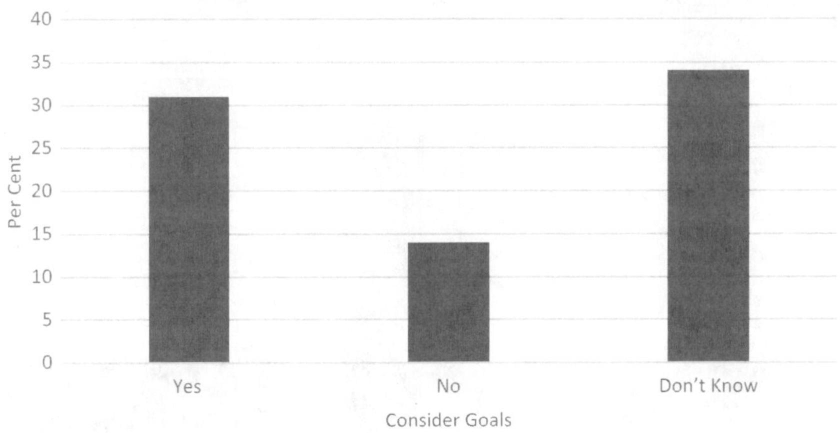

Figure 4.10 Consider County Council Goals.

Reversing the direction, members of town and parish councils were asked to indicate whether they, as a parish council, considered the goals of the principal authority. Figure 4.10 shows that 31 per cent said that they did consider the goals of the principal authority; 14 per cent said that they did not consider the goals of the principal authority; and 34 per cent did not know.

Responses to these questions suggest that at the time of data collection there was a disconnect between the expectations of parish council and the county council. Data was collected during a time of transition, and this inevitably raised concerns about the effectiveness of future governance structures. The high level of non-response relating to formal linkage suggests uncertainty and disengagement, and the majority of perceptions that the county council is not adequately considering parish needs reflects broader concerns about the marginalisation of the parish tier in England (Copus et al. 2017). It has also been recognised that principal authorities may prioritise strategic service delivery over the perceived need to parish councils, creating a structural tension (Sandford 2017). The results presented here also reflect on recent observations that over a fifth of local councils felt disconnected from county council tiers of government. As the future of local councils report observes, whilst '...some councils are actively engaged with other tiers of local government and other local councils...a worryingly large number miss crucial broader engagement' (Giovannini et al. 2023).

Whilst there is no scope to explore this in any detail here it is worth summarising general impressions gained from within interview data. Inevitably there was a diverse range of views on how a unitary authority would engage and support parish councils. And whilst this appeared to be predominantly negative from the survey data, the interviews presented a much more complex and nuanced understanding of the potential impact of the unitary authority. The role of partnerships formed a central theme within the research.

Partnership Expanded

Beyond connections to the county council, respondents were asked to indicate to what extent they wanted to and did work in partnership with other organisations.

Respondents were asked if they felt that it was important to work in partnership. Figure 4.11 shows that, overwhelmingly, at 96 per cent, respondents indicated that working in partnership was important. Only 0.5 per cent said that working in partnership was not important, and 1 per cent said that they did not know if working in partnership was important.

Respondents were asked to indicate whether they felt that it was important to work in partnership with a number of other organisations. Members of town and parish councils were also asked to indicate whether they felt that they were actually in partnership with the organisations presented. Firstly, respondents were asked to indicate whether partnership was important with national government.

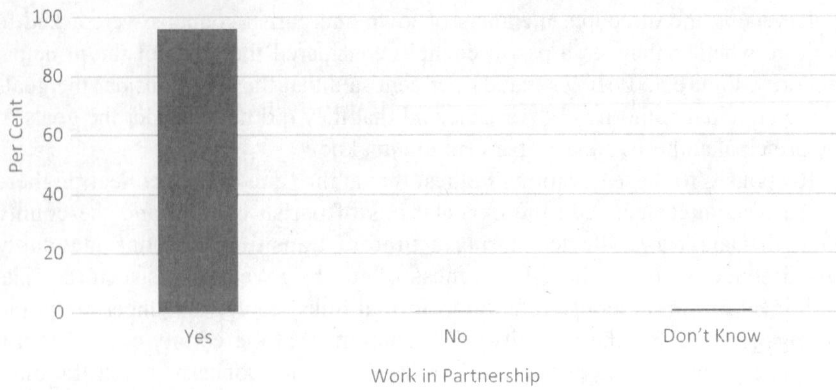

Figure 4.11 Work in Partnership.

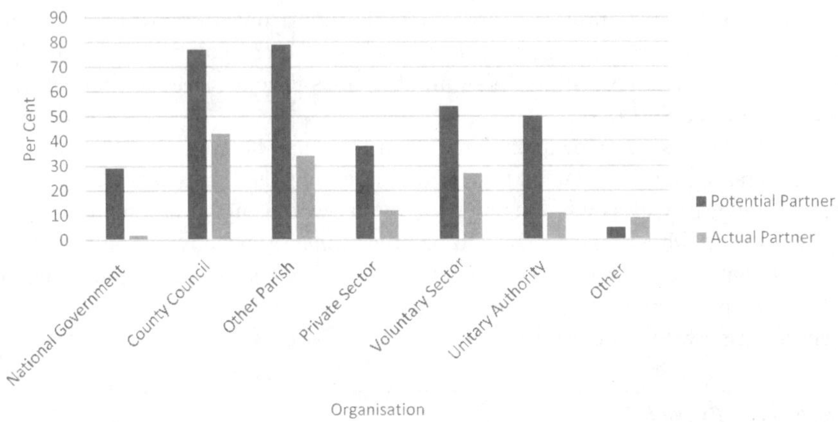

Figure 4.12 Partners.

Figure 4.12 shows that 29 per cent of respondents felt that it was important to be in partnership with national government, but only 2 per cent felt that they were actually in partnership with national government. Secondly, and building on previous results, respondents were asked to indicate whether partnership was important with their county council. The large majority, at 77 per cent, indicated that 'yes' working in partnership with their county council was important. However, when asked to indicate whether they felt that they were actually in partnership with their county council, only 43 per cent indicated that this was actually in place. Thirdly, participants in the survey were asked to indicate whether it was important to work in partnership with other parishes; 79 per cent indicated that they thought it was important to work in partnership with other parishes; and 34 per cent said that they were actually in partnership with other parishes. Fourthly, respondents were asked to indicate whether they thought it was important to work in partnership with the private sector; 38 per cent indicated that it was important to work in partnership with the private sector; and 12 per cent indicated that they were actually in partnership with the private sector.

Acknowledging that members of parish councils are themselves volunteers, participants were asked to indicate if they felt that it was important to be in partnership with the voluntary sector; 54 per cent said that it was important to work in partnership with the voluntary sector; and 27 per cent indicated that they were actually in partnership with the voluntary sector. Finally, respondents were asked to indicate whether they felt it was important to work in partnership with a unitary authority that was in the process of being established at that time; 50 per cent of the respondents indicated that it was important to work in partnership with a unitary authority; and only 11 per cent felt that they were actually in partnership with a unitary authority.

Overall respondents indicated that they felt working in partnership was important to the operational success of their parish councils. By extension these results also provide insights into the role of partnerships in community coherence more generally. Local government in England is increasingly framed within the tension of place-based leadership and networked governance, a process, indeed our previous discussions have explicitly drawn out these tensions. This brings to the fore the importance of cross-sectoral partnerships in multiple areas, from service delivery to embedding sustainable development policies in local communities. However, results presented here demonstrate a significant gap between the perceived importance of partnerships and respondents' actual experience of working in partnership. These observations resonate with broader critiques of partnership that point to rhetoric over substance. It is recognised that partnerships with other tiers of government can be asymmetrical, with an imposed agenda instead of genuine consultation and local engagement. However, greater interest in partnerships with other parish councils and the voluntary sector supports observations that the more localised and horizontal partnerships are perceived as being more directly relevant to local needs. These findings are also contextualised by the discussion in Chapter 3 relating to broader policy and legislative frameworks.

Barriers to Service Provision

Responsibility for service provision at local government level varies between the statutory responsibility of the principal authorities and that of the parish council. In Cornwall, as in other unitary authorities in England, statutory services, such as education, planning, social care, highways, and waste management, are delivered by Cornwall Council. Parish councils focus on local amenities such as allotments, cemeteries, community centres, parks, and have responsibility for neighbourhood planning. As we have already seen, the parish council can be the first point of contact for local residents and play an essential role in communicating concerns and local needs to the principal authority. Ultimately the effectiveness of service provision depends on the quality of collaboration and communication between these tiers of government. Therefore, extending the previous analysis and making connections between governance and operational capacity, the survey sought to explore the potential barriers to service provision within the parish area. The focus here was not on the quality of communication between the county council; this has already been addressed within the context of partnership. Instead councillors were presented with a number of issues unique to Cornwall that may impact service provision. This included the scattered layout of Cornish communities, lack of involvement in the planning process, an ageing population, and the complex needs of Cornish communities.

Respondents were asked to indicate on a scale of 1 to 5, whether the barrier outlined was very significant, 1, or alternatively, not at all significant, 5.

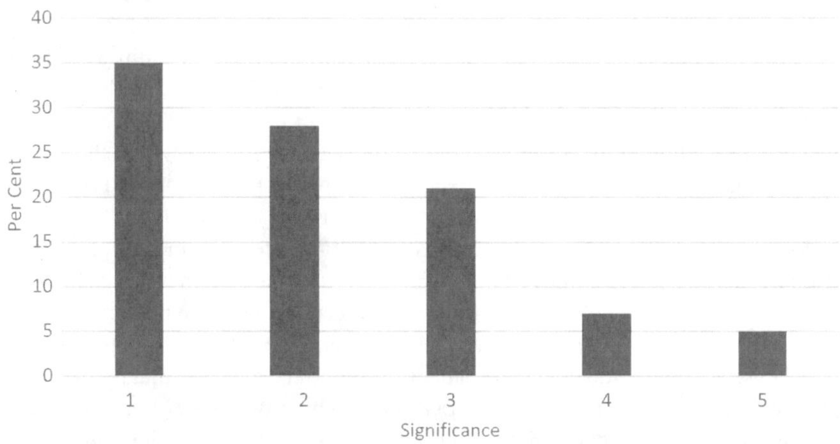

Figure 4.13 Scattered Layout of Communities.

As we have already discussed, Cornwall, as a predominantly rural county, has a diverse and scattered built environment. Settlements range from small hamlets such as Doddycross in Southeast Cornwall with approximately 50 residents to towns such as St Austell that have over 27,000 residents. Members of town and parish councils were asked to indicate whether they felt that the scattered layout of Cornwall's communities was a significant barrier to effective service delivery. Figure 4.13 shows that 35 per cent stated that scattered communities was a very significant barrier to service delivery; 28 per cent indicated that scattered communities was a significant barrier to service delivery; 21 per cent indicated that scattered communities was a moderately significant barrier to service delivery; 7 per cent stated that the scattered nature of Cornwall's communities was of little significance; and 5 per cent said that the scattered nature of Cornwall's communities was not significant to the adequate provision of services.

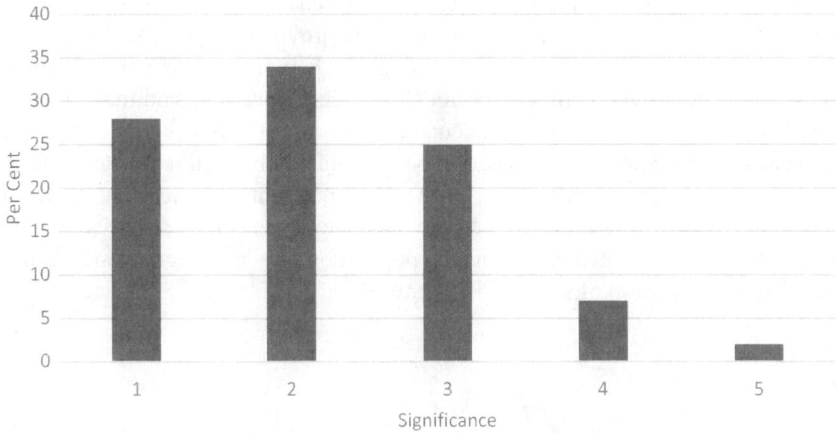

Figure 4.14 Lack of Involvement.

Another identified barrier to service provision was the lack of involvement in the planning process. Members of town and parish councils were asked to indicate to what degree they felt that a lack of involvement of community and individuals in the planning, development, and delivery of services provides a barrier to adequate service provision in Cornwall. Figure 4.14 shows that 28 per cent indicated that a lack of involvement was very significant; 34 per cent said that a lack of involvement was significant; 25 per cent indicated that a lack of involvement was moderately significant; 7 per cent stated that a lack of involvement was of little significance; and 2 per cent said that a lack of involvement was not of significant at all.

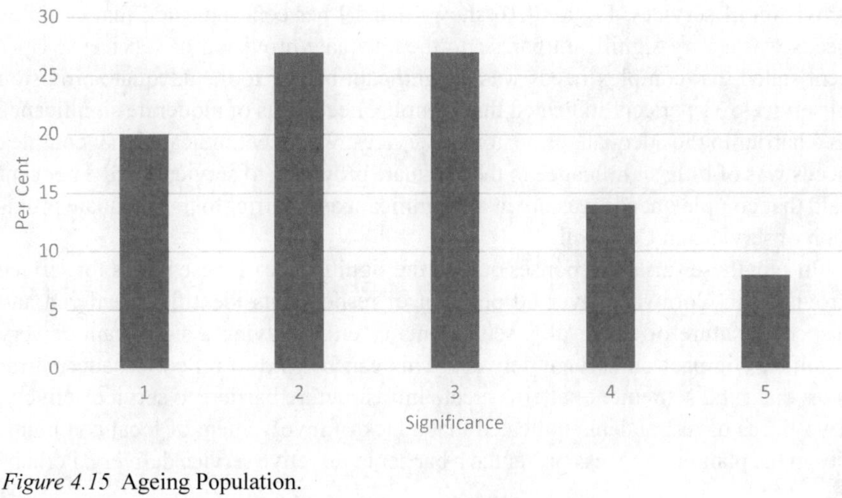

Figure 4.15 Ageing Population.

Members of town and parish councils were asked to indicate how significant an ageing population of Cornwall is to the adequate provision of services. Figure 4.15 shows that 20 per cent indicated that an ageing population was very significant to the adequate provision of services in Cornwall; 27 per cent said that an ageing population was significant to the adequate provision of services in Cornwall; 27 per cent said that an ageing population was a moderately significant barrier to the adequate provision of services in Cornwall; 14 per cent indicated that an ageing population is of little significance to the adequate provision of services in Cornwall; and 8 per cent stated that an ageing population was of no significance at all to the adequate provision of services in Cornwall.

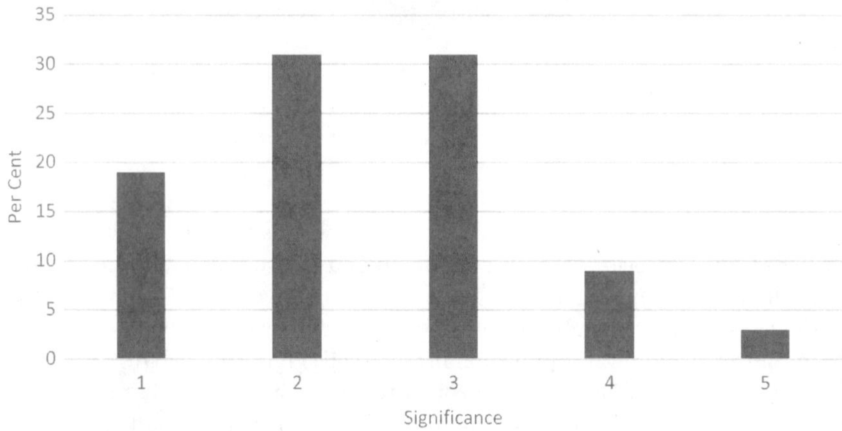

Figure 4.16 Complex Needs.

Members of town and parish councils were asked to indicate how significant the complex needs of Cornwall's communities are as a barrier to the adequate provision of services. Figure 4.16 shows that 19 per cent indicated that complex needs was a very significant barrier to the adequate provision of services; 31 per cent stated that complex needs was a significant barrier to the adequate provision of services; 31 per cent indicated that complex needs was of moderate significance as a barrier to the adequate provision of services; 9 per cent indicated that complex needs was of little significance to the adequate provision of services; and 3 per cent said that complex needs was not at all significant as a barrier to the adequate provision of services in Cornwall.

In sum these survey responses outline the significance of these issues for service provision in Cornwall. Over 60 per cent of respondents identified scattered and dispersed nature of Cornwall's settlements as either having a significant or very significant impact on service delivery. This can be said to reflect broader narratives that rural settlement patterns create infrastructure barriers to service delivery. Two-thirds of respondents indicated that a lack of involvement of local communities in the planning process presented a barrier to effective service delivery. Perhaps this represents border voices that rural areas are in need of stronger participatory approaches.

Conclusion

This chapter has presented an overview of the research area, Cornwall. This chapter has focused on Cornwall's unique cultural and governance context. We have briefly identified the key challenges and opportunities that Cornwall faces. We have also seen some of the main changes in governance structures from the time the data was collected to today. This was predominantly related to the transition of Cornwall to a unitary authority. The methodological approach taken has been discussed, and I have introduced the core reflexivity indicators that underpin the relationship between sustainable development and reflexive modernity from an analytical point of view. Key statistical data, with some relevant interview excerpts, have further provided the geographical, demographic, and governance context of the research. Chapter 5 builds on this foundation, examining how Cornwall's local councils understand and articulate sustainable development through their own voices and stories.

Bibliography

Allmendinger, P., & Haughton, G. (2010). Spatial planning, devolution, and new planning spaces. *Environment and Planning C: Government and Policy*, 28(5), 803–818.

Bamberg, S., & Möser, G. (2007). Twenty years after Hines, Hungerford, and Tomera: A new meta-analysis of psycho-social determinants of pro-environmental behaviour. *Journal of Environmental Psychology*, *27*(1), 14–25.

Borne, G. J. (2006) Sustainable development: the reflexive governance of risk (Thesis, University of Plymouth).

Braun, V., & Clarke, V. (2021). *Thematic analysis: A practical guide*. Sage.

Bryant, A., & Charmaz, K. (2007). Grounded theory in historical perspective: An epistemological account. In Bryant, A. and Charmaz, K. (Eds.), *The SAGE handbook of grounded theory* (pp. 31–57). Sage.

Bucholtz, M. (2000). The politics of transcription. *Journal of Pragmatics*, *32*(10), 1439–1465.

Charmaz, K. (2014). *Constructing grounded theory*. SAGE.

Charmaz, K. (2020). "With constructivist grounded theory you can't hide": Social justice research and critical inquiry in the public sphere. *Qualitative Inquiry*, *26*(2), 165–176.

Clarke, A. (Ed.). (2005). *Situational analysis: Grounded theory after the postmodern turn*. Sage Publications.

Copus, C., Roberts, M., & Wall, R. (2017). *Local government in England: Centralisation, autonomy and control*. Springer.

Corbin, J., and Strauss, A. (1990). Grounded Theory Research: Procedures, Canons and Evaluative Criteria. *Qualitative Sociology*, 13(1), 3–21

Cornwall Council. (2025). The Cornwall we know 2025: Evidence and intelligence. Cornwall Councilhttps://letstalk.cornwall.gov.uk/the-cornwall-we-know

Cornwall Council. (2023). Community area partnerships: New approach to local engagement and decision-making. https://www.cornwall.gov.uk/people-and-communities/community-area-partnerships/

Cornwall and Isles of Scilly Local Enterprise Partnership (LEP). (2022). Vision 2030: Clean and inclusive economic growth. Retrieved from https://www.cioslep.com

Council of Europe. (2025). *Framework convention for the protection of national minorities*. Council of Europe. Retrieved from https://www.coe.int/en/web/minorities

Department for Levelling Up, Housing and Communities (DLUHC). (2023). Cornwall devolution deal (Kevambos Digresennans Kernow) 2023. GOV.UK. https://www.gov.uk/government/publications/cornwall-devolution-deal-kevambos-digresennans-kernow-2023

Department for Environment, Food & Rural Affairs (DEFRA). (2019). Public attitudes tracker: Wave 30 – December 2019. GOV.UK. www.gov.uk/government/statistics/beis-public-attitudes-tracker-wave-30

Dey, I. (1996). *Qualitative data analysis: A user friendly guide for social scientists*. Routledge.

European Commission. (2019). Special Eurobarometer 501: Attitudes of European citizens towards the environment. Publications Office of the European Union. https://europa.eu/eurobarometer

Fairclough, N. (2003). *Analyzing discourse: Textual analysis for social research*. Routledge.

Gallent, N. (2014). The social value of second homes in rural communities. *Housing, Theory and Society*, *31*(2), 174–191.

Giovannini, A., Griggs, S., Jones, A., Parker, S., & Rose, J. (2023). *The future of local councils*. Society for Local Council Clerks (SLCC).

Hajer, M. (1995). *The politics of environmental discourse: Ecological modernisation and the policy process*. Oxford University Press.

Holstein, J., & Gubrium, J. (1994). Phenomenology, ethnography and interpretative practice. In N. Denzin & S. Lincoln (Eds.), *Handbook of qualitative research* (pp. 262–272). Sage.

Kvale, S. (1996). *Interviews: An introduction to qualitative research interviewing*. Sage Publications.

Leal Filho, W., Raath, S., Lazzarini, B., Vargas, V. R., de Souza, L., Anholon, R., … & Orlovic, V. L. (2018). The role of transformation in learning and education for sustainability. *Journal of Cleaner Production*, *199*, 286–295.

Lowndes, V., & Sullivan, H. (2004). Like a horse and carriage or a fish on a bicycle: How well do local partnerships and public participation go together? *Local Government Studies*, *30*(1), 51–73.

National Association of Local Councils. (2025). National association of local councils. https://www.nalc.gov.uk

Nikander, P. (2008). Working with transcripts and translated data. *Qualitative Research in Psychology*, *5*(3), 225–231.

O'neill, S., & Nicholson-Cole, S. (2009). "Fear won't do it" promoting positive engagement with climate change through visual and iconic representations. *Science Communication*, *30*(3), 355–379.

Otto, S., & Pensini, P. (2017). Nature-based environmental education of children: Environmental knowledge and connectedness to nature, together, are related to ecological behaviour. *Global Environmental Change*, *47*, 88–94.

Poortinga, W., Whitmarsh, L., Steg, L., Böhm, G., & Fisher, S. (2019). Climate change perceptions and their individual-level determinants: A cross-European analysis. *Global Environmental Change*, *55*, 25–35.

Sandford, M. (2017). *Local government in England: Structures*. House Commons Libr London.

Shucksmith, M. (2018). Re-imagining the rural: From rural idyll to Good Countryside. *Journal of Rural Studies*, *59*, 163–172.
Silverman, D. (1998). Qualitative/quantitative. In C. Jenks (Ed.), *Core sociological dichotomies* (pp. 78–95). Sage.
Society of Local Council Clerks. (2024). Society of local council clerks. https://www.slcc.co.uk
Thornberg, R. (2012). Informed grounded theory. *Scandinavian Journal of Educational Research*, *56*(3), 243–259.
Thornberg, R., & Dunne, C. (2019). Literature review in grounded theory. *The Sage Handbook of Current Developments in Grounded Theory*, *2*, 206–221.
UK Government. (2014, April 24). Cornish granted minority status within the UK. GOV.UK. www.gov.uk/government/news/cornish-granted-minority-status-within-the-uk
Van Dijk, T. A. (2017). *Discourse and power*. Bloomsbury Publishing.
Whitmarsh, L., & O'Neill, S. (2010). Green identity, green living? The role of pro-environmental self-identity in determining consistency across diverse pro-environmental behaviours. *Journal of Environmental Psychology*, *30*(3), 305–314.
Whitmarsh, L., O'Neill, S., & Lorenzoni, I. (2011). Climate change or social change? Debate within, amongst, and beyond disciplines. *Environment and Planning A*, *43*(2), 258–261.
Willett, J. (2023a). Counter-urbanisation and a politics of place: A coastal community in Cornwall and rural gentrification. *Habitat International*, *141*, 102935.
Willett, J. (2023b). Place-based rural development: A role for complex adaptive region assemblages? *Journal of Rural Studies*, *97*, 583–590.
Wodak, R., & Meyer, M. (Eds.). (2015). *Methods of critical discourse studies*. Sage.
Woods, M. (2010). *Rural*. Routledge.

5 Sustainable Development and Risk

Initially the concept of sustainable development is examined in the context of its visibility and salience to town and parish councils. The form that sustainable development takes is then discussed using the three pillars approach to identify priorities and trade-offs. The abstract interpretation of the concept is related directly to how relevant councillors feel sustainable development is to Cornwall. To that end a number of statements designed to further examine perspectives on sustainable development are discussed.

Discourses of sustainable development are then related to local governance mechanisms, identified as having the ability to embed sustainable development in local governance structures. As this chapter unfolds we begin to explore perceptions and engagement with issues and risks that are associated with sustainable development. Global warming is identified as a prominent risk associated with sustainable development, both at the local and international levels. Engagement with global warming is explored, and this is related directly to councils' interpretation of global warming within Cornwall. There is then a continued focus on potential drivers for climate change within Cornwall, making tangible connections between the global and local scales of analysis. The final section of this chapter focuses very directly on the relationship between the global and local, presenting survey results on how respondents prioritise risks associated with sustainable development.

Building on the synergies that were outlined in Chapter 2, which establishes the theoretical connection of sustainable development with a reflexive modernity, and as we progressively explore perceptions and engagement with sustainable development as well as associated issues, it is possible to address all of the reflexive modern indicators identified in Table 5.1, first introduced in chapter 4. This is achieved in the survey by progressively building on engagement and perceptions of sustainable development, which is based on identifying to what extent respondents understand sustainable development in relation to the three pillars. Using this as a foundation different statements about what sustainable development may mean are presented to respondents in order to understand how councillors identify trade-offs between society, economy, and the environment.

DOI: 10.4324/9780429297199-5

Table 5.1 Indicators of Reflexive Modernity

Indicator of Reflexive Modernity	*Questionnaire*
1 Displacement of traditional authority	Reveals how much governance is devolved and how councils work with other institutions – signs of a shift from top-down to networked decision-making. Elements of parish planning, partnerships, and strategic goals.
2 Knowledge pluralism/critique of expert systems	Highlights local views on whether local authorities understand or consider local needs. Identify resistance to imposed expert frameworks.
3 Ethical and moral reasoning	Explores whether sustainable development is framed in moral terms such as intergenerational equity or dismissed as jargon. Represents reflexive moral engagement or scepticism.
4 Sustainable consumption/reflexive choice-making	Captures personal and parish-level practices like recycling, awareness of carbon footprint, and behavioural change. Representative of reflexive environmental subjectivity.
5 Reflexive governance structures	Assesses whether councils use adaptive processes, for example, feedback from the community and work in evolving partnerships. Representative of experimental, reflexive governance.
6 Global-local risk awareness	Reveals whether global risks like climate change are seen as relevant to Cornwall linking local sustainable development actions to global reflexive concerns.

Translating these perceptions of sustainable development into an understanding of local governance mechanisms for achieving sustainable development, respondents are then asked to identify their understanding of, and engagement with, key vehicles identified for embedded sustainable development in the local areas. This included Local Areas Agreements, Comprehensive Area Assessments, Sustainable Community Strategy, and Quality Parish. The survey then moves forward to explore perceptions of risks associated with sustainable development. As previously mentioned, there is a focus on global warming as a risk that is a dominant narrative in sustainable development discourse. And again the abstract perceptions are grounded through questions in the survey that ask respondents to identify the extent to which certain issues within Cornwall are drivers for global warming. Finally councillors are asked to make global and local connections through the salience of particular risks. Together these elements of sustainable development allow us to explore reflexive modern dynamics within parish councils and their communities.

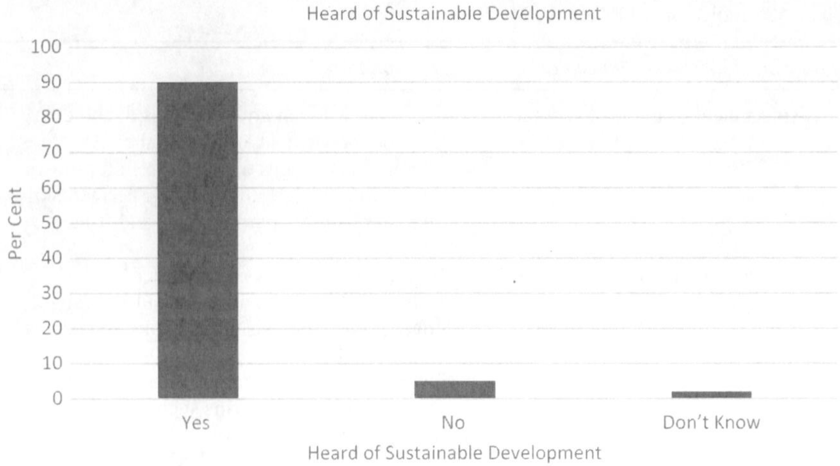

Figure 5.1 Sustainable Development

Members of town and parish councils were asked to indicate if they had heard of sustainable development. Figure 5.1 shows that the vast majority, at 90 per cent, said that they had heard of sustainable development; 5 per cent said that they had not heard of sustainable development; 2 per cent said that they did not know if they had heard of sustainable development.

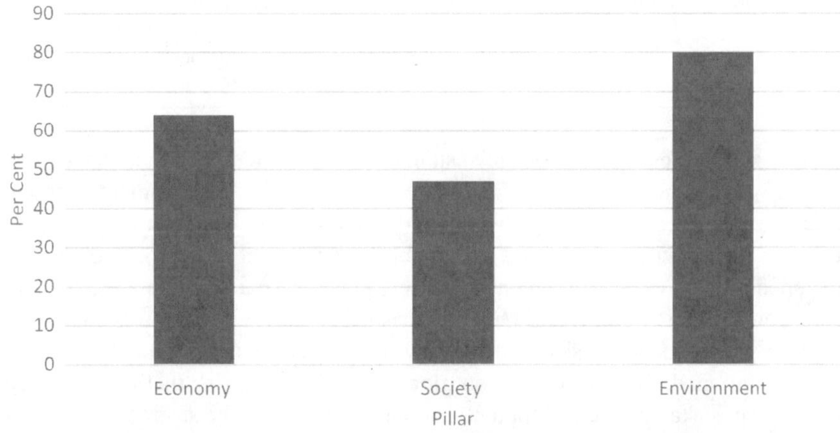

Figure 5.2 Three Pillars

Respondents were asked to indicate which of the three pillars of economy, society, and environment they thought sustainable development referred to. Figure 5.2 shows that, on aggregate, 64 per cent said that sustainable development was the economy, and 47 per cent said that sustainable development was society. The majority of respondents, at 80 per cent, indicated sustainable development referred to the environment.

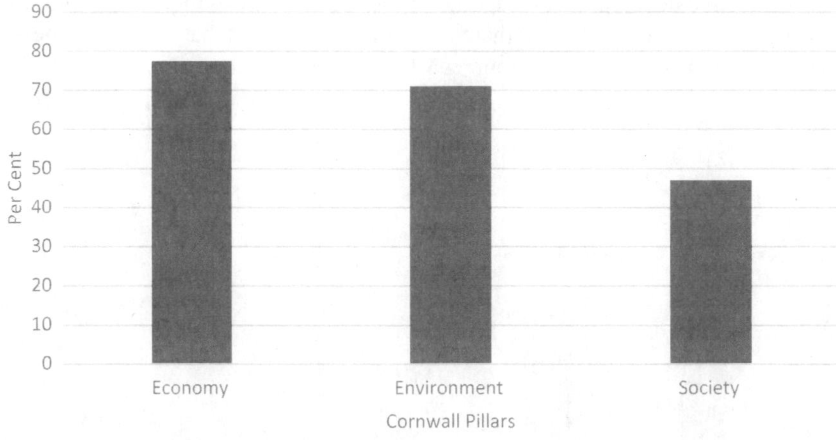

Figure 5.3 Cornwall Pillars

Members of town and parish councils were also asked if they would indicate which of the three pillars of sustainable development were important in Cornwall. Figure 5.3 shows that 77.5 per cent indicated that the economy was most important to Cornwall; 71 per cent said that the environment was most important to Cornwall; and 47 per cent of respondents indicated that society was important to Cornwall.

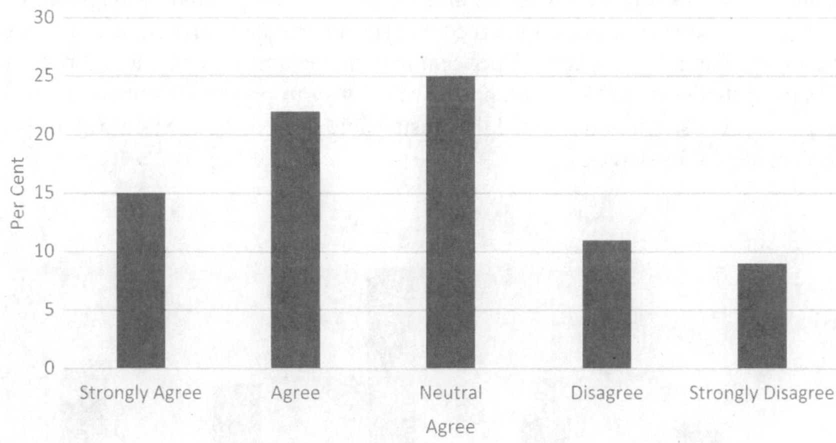

Figure 5.4 Environment More Important Than Money

Members of town and parish councils were asked to indicate if they felt that sustainable development meant that the environment was more important than money. Figure 5.4 shows that 15 per cent strongly agreed with this statement; 22 per cent agreed that sustainable development meant that the environment was more important than money; 25 per cent indicated that they were neutral; 11 per cent disagreed that sustainable development meant that the environment was more important than money; and 9 per cent indicated that they strongly disagreed that sustainable development meant that that the environment was more important than money.

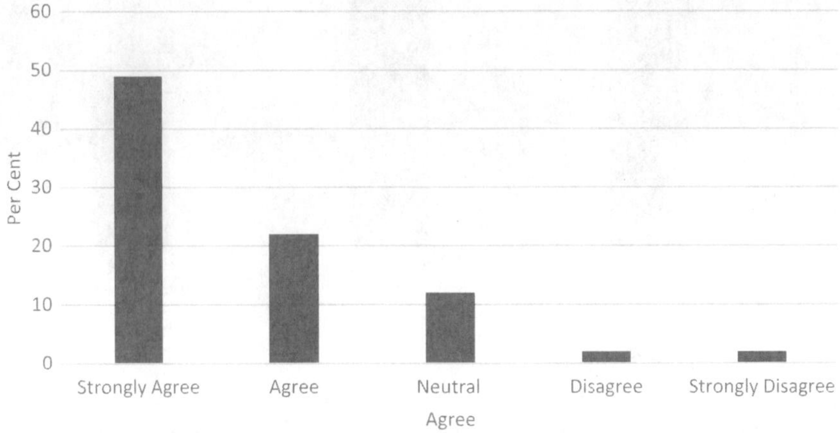

Figure 5.5 Preserve Natural Resources

Members of town and parish councils were asked to indicate to what extent they agreed with the statement that sustainable development is about preserving natural resources. Figure 5.5 shows that 49 per cent said that they strongly agreed with this statement; 22 per cent agreed that sustainable development was a way of preserving natural resources; 12 per cent maintained a neutral position; 2 per cent disagreed that sustainable development was a way to preserve natural resources; and 2 per cent strongly disagreed that sustainable development was a way of preserving natural resources.

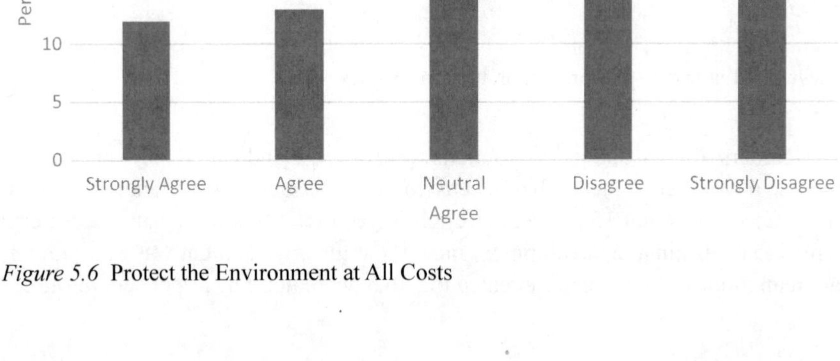

Figure 5.6 Protect the Environment at All Costs

Members of town and parish councils were asked to indicate to what extent they agreed with the statement that sustainable development was a term that meant the protection of the environment at all costs. Figure 5.6 shows that 12 per cent of respondents strongly agreed with this statement; 13 per cent indicated that they agreed that sustainable development meant the protection of the environment at all costs; 26 per cent indicated a neutral position on this statement; 16 per cent disagreed that sustainable development meant the protection of the environment at all costs; and 17 per cent strongly disagreed that sustainable development meant the protection of the environment at all costs.

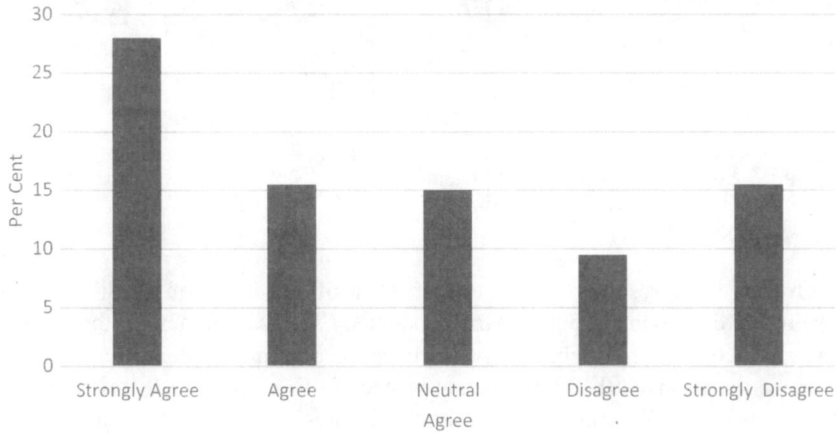

Figure 5.7 Sustainable Development Is Jargon

Members of town and parish councils were asked to indicate to what extent they agreed with the statement that sustainable development was jargon to make governments sound greener. Figure 5.7 shows that 28 per cent strongly agreed with this statement; 15.5 per cent agree that sustainable development is jargon designed to make governments sound greener; 15 per cent indicated a neutral position concerning this statement; 9.5 per cent disagreed that sustainable development was jargon to make governments sound greener; and 15.5 per cent strongly disagreed that sustainable development is jargon to make governments sound greener.

Figure 5.8 represents an aggregate of responses that indicated that they strongly agreed or agreed with the aforementioned statements. The majority of respondents at 71 per cent indicated that they strongly agreed or agreed that sustainable development was a term that meant preserving natural resources; 43.5 per cent indicated that they either agreed or strongly agreed that sustainable development was jargon; 37 per cent said that sustainable development meant that the environment was more important than money; and 25 per cent indicated that sustainable development meant that the environment should be protected at all costs.

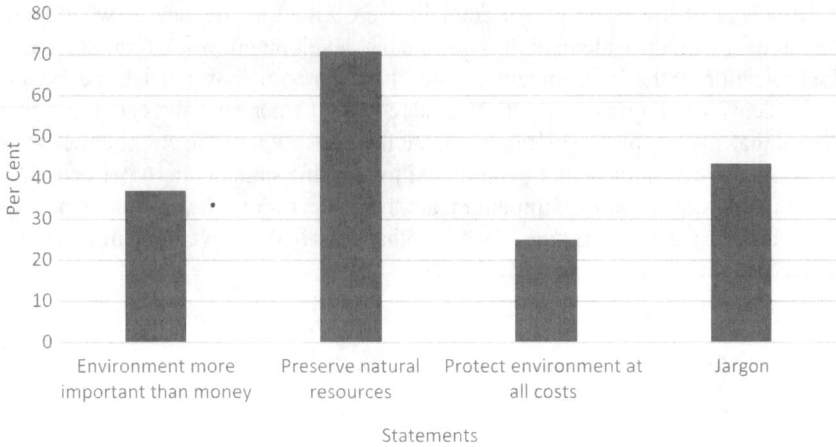

Figure 5.8 Sustainable Development Statements

Overall, these results reveal both the extent of and limits of the discourse of sustainable development within parish councils. Clearly awareness of the concept was reported as being high, and yet its meaning was inconsistent, ranging from strong associations with environmental protection to a complete dismissal of the concept as jargon. We can reengage with the debates in Chapter 2 in regard to criticism of the concept, where we have already established the term as ambiguous and contested. Specifically, councillors' emphasis on environmental issues alongside the prioritisation of the economy in Cornwall reflects the way that the three pillars are unevenly understood in practice. From the reflexive modernity perspective the findings present patterns of partial reflexivity. What seems clear is that respondents are interrogating the authority of the abstracted expert-led discourses, relating them to their own lived experiences (Beck et al. 2003; Fischer 2000). Furthermore, the prioritisation of economic concerns within Cornwall could point to a form of reflexive adaptation of global sustainable development discourse, where three pillars are rebalanced to accommodate local conditions and the perceived relevance of risks (Purvis et al. 2019). Similarly, the ambivalence expressed towards global warming and the uneven recognition of the social dimension point towards a reflexivity where councillors are negotiating manufactured risks and questioning governance processes. In light of this analysis what I suggest is that the partial and contested interpretations of sustainable development do not undermine reflexivity but instead represent it. What we witness is a glimpse at how global discourses are mediated, resisted, and then reconfigured through the everyday practice of parish governance.

Sustainable Development Mechanisms

Having established broad perceptions of sustainable development, respondents were asked to indicate if they were aware of any of the mechanisms that existed within the local government arena that can be said to promote sustainable communities and sustainable development. As discussed in Chapter 3, at the time the research was conducted, frameworks for examining sustainable development were underdeveloped and fragmented. It was therefore necessary to identify strategies, programmes, and initiatives that fed into achieving aspect of sustainable development at the local level. In the context of local government, these mechanisms were identified as Local Area Agreements, Community Area Agreements, Sustainable Community Strategy, and Quality Parish status.

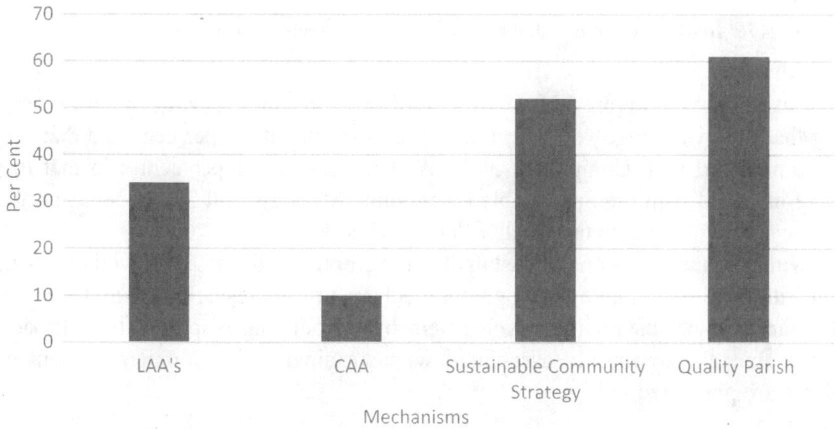

Figure 5.9 Awareness of Sustainable Development Mechanisms

Figure 5.9 shows that 34 per cent said that they had heard of Local Area Agreements; 10 per cent said that they had heard of Comprehensive Area Assessments; 52 per cent of respondents said that they were aware of the Sustainable Community Strategy; 61 per cent indicated that they were aware of the Quality Parish scheme. On the whole, interviews confirmed the survey findings. There was an overall low level of understanding of the mechanisms that could be utilised to promote sustainable development and sustainable communities more broadly. Building on awareness of identified sustainable development mechanisms, respondents were also asked to indicate to what degree they felt that they were involved with these mechanisms.

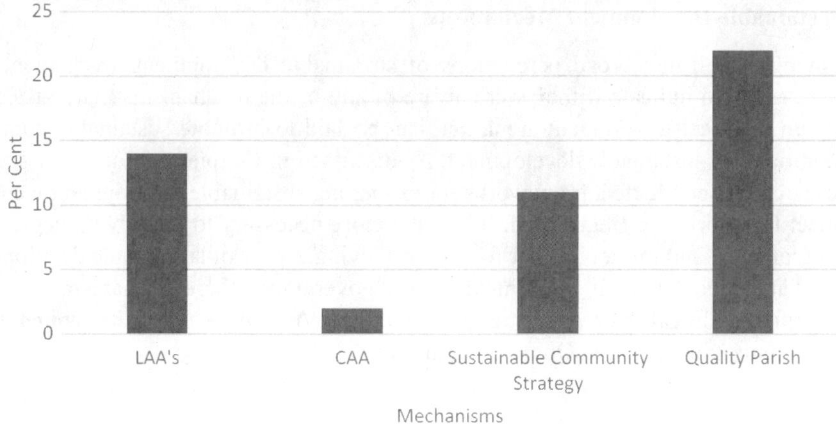

Figure 5.10 Involved with Sustainable Development Mechanisms

Unsurprisingly, Figure 5.10 reinforces the previous findings with 14 per cent saying that they were involved with Local Area Agreements; 2 per cent said that they were involved with Comprehensive Area Assessments; 11 per cent said that they were involved with the Sustainable Community Strategy; and 22 per cent said that they were involved with the Quality Parish Scheme.

With initial perceptions of sustainable development discussed as well as engaging with local sustainable development mechanisms, the research sought to expand the narrative of sustainable development by introducing respondents to broader issues that are commonly associated with sustainable development discourses. These are presented in Figure 5.11.

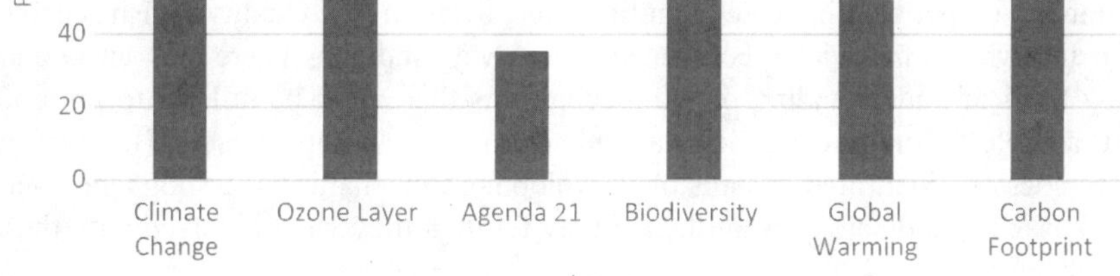

Figure 5.11 Aware Sustainable Development Related Issues

Respondents were presented with a list of issues that related to sustainable development and asked to indicate if they were aware of a particular issue. Figure 5.11 shows that on aggregate 98 per cent said that they had heard of climate change; 87 per cent said that they had heard of the ozone layer; 35 per cent had heard of Agenda 21; 84.5 per cent said that they had heard of biodiversity; 98 per cent said that they had heard of global warming; and 95.5 per cent said that they had heard of carbon footprint.

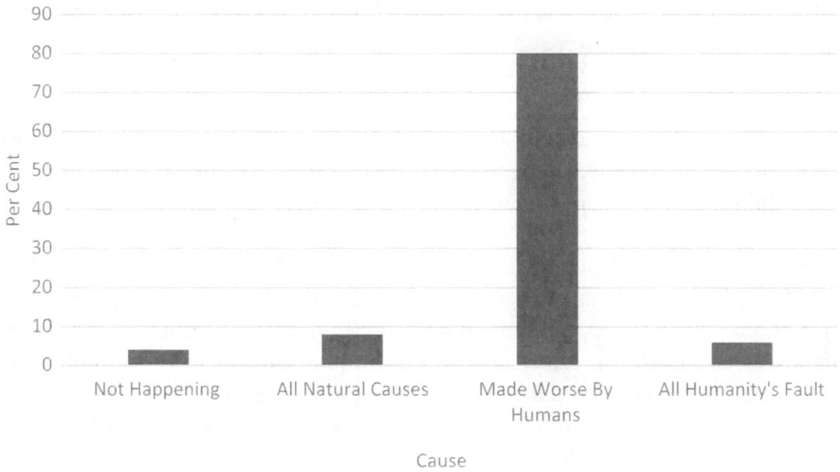

Figure 5.12 Global Warming

In the survey global, warming as a distinct term was initially used instead of climate change or global climate change in order to reduce confusion with stratospheric ozone depletion and the greenhouse effect (Stamm et al. 2000; Wilson 2000). Moreover, as Whitmarsh (2009) indicates, the term global warming was seen to evoke more concern than the term climate change. As such the term global warming is used throughout the research and this book. As is evident in Figure 5.12, the survey presented respondents with four options as to the extent of humanity's impact on the processes of global warming. Four per cent suggested that global warming was not really happening; 8 per cent indicated that global warming was a natural occurrence; overwhelmingly, 80 per cent agreed that global warming was made worse by humans; and 6 per cent agreed that global warming was all of humanity's fault.

Whilst we will elaborate on this point in more detail in subsequent chapters, the central role that climate change and global warming play in sustainable development discourse warrants a deeper examination at this point. Results from the interviews were very varied as to the perceived cause-and-effect relationship of global climate change. Many interviewees agreed it was an important issue and 'something should be done'. The following extract typifies this sort of response:

> After the winter we've just had I wonder, but certainly as far as global warming goes I think it should concern everybody and we should all be trying to save. I mean I look at council buildings, government buildings and they're flooded with lights, in broad daylight, why do they need all the lights on, if they tried to ensure that 50% of the lights were put on and 50% were switched off, in other words perhaps in a office it may need rewiring but the point is that if 50% were saved on every government building how much would you save?

As illustrated in the statement above the interviewee emphasizes to the importance of climate change, indicating that it should be a high priority. However, the opening of this statement, 'after the winter we've had', pinpoints the subtle undertones of uncertainty and scepticism. This is a finding echoed in broader research that looks at public identification with climate change (Hulme 2009, 2023). With the previous assessment in mind many interviewees were unclear as to the types of actions that could be taken to adapt to or mitigate what were often seen as ambiguous effects. There was a significant narrative within the data suggesting that the issues that surround global warming were not pertinent to the work of parish councils, something we will also explore in subsequent chapters. When asked about thoughts on global warming the sentiment contained in the following statement was not uncommon:

> I don't, quite honestly, I don't understand it. I watched a programme the other night I'm fascinated with polar bears or bears of any sort

Nevertheless the political and social landscape of the time, and magnified in subsequent years, has been dominated by the spectre of climate change and the impacts and challenges that this global phenomenon presents humanity with. And indeed, the gravity of this global risk is such that in his last book, Ulrich Beck explores climate change in the context of a risk society, speculating on the potential emancipatory qualities climate change offers (Beck 2016). And so, we move forward in this chapter with a stronger focus on global climate change at the local level, asking survey respondents to indicate not only whether they felt that global warming was important to Cornwall, but also what Cornwall's specific attributes are drivers for global warming in the county.

Cornwall and Global Warming

Members of town and parish councils were asked to indicate whether they felt that global warming was an important issue for Cornwall. Figure 5.13 shows that the

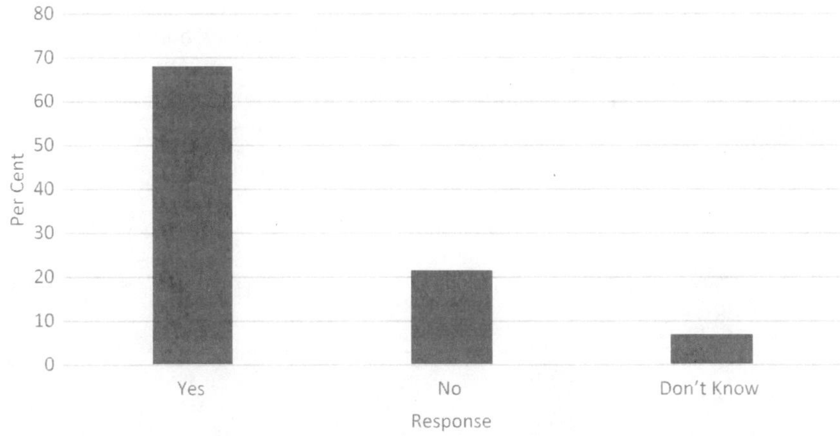

Figure 5.13 Global Warming and Cornwall

majority of respondents, at 68 per cent, said that global warming was an important issue for Cornwall; 21.5 per cent indicated that global warming was not an important issue for Cornwall; and 7 per cent said that they did not know if global warming was an important issue for Cornwall. Building on this initial observation members of town and parish councils were presented with a number of factors identified by Cornwall Council as drivers for climate change in Cornwall.

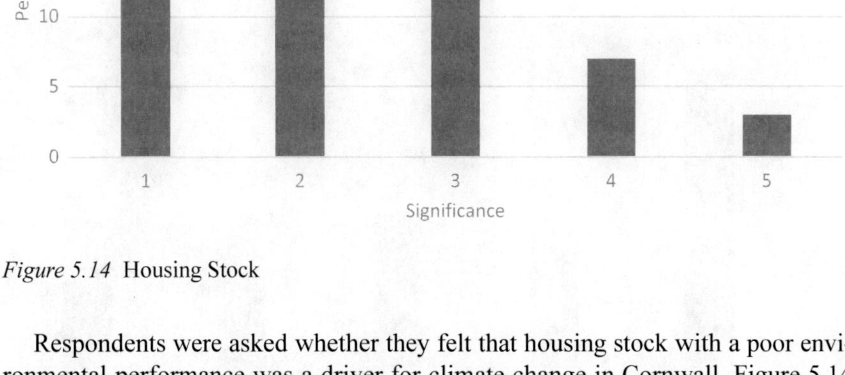

Figure 5.14 Housing Stock

Respondents were asked whether they felt that housing stock with a poor environmental performance was a driver for climate change in Cornwall. Figure 5.14 shows that 19 per cent indicated that a poor housing stock was a very significant driver for change; 21 per cent indicated that a poor housing stock was a significant driver for climate change; 20 per cent indicated that a poor housing stock was moderately significant as a driver for climate change in Cornwall; 7 per cent said that poor housing stock was of little significance as a driver for climate change in Cornwall; and 3 per cent said that housing stock was not at all significant as a driver for climate change in Cornwall.

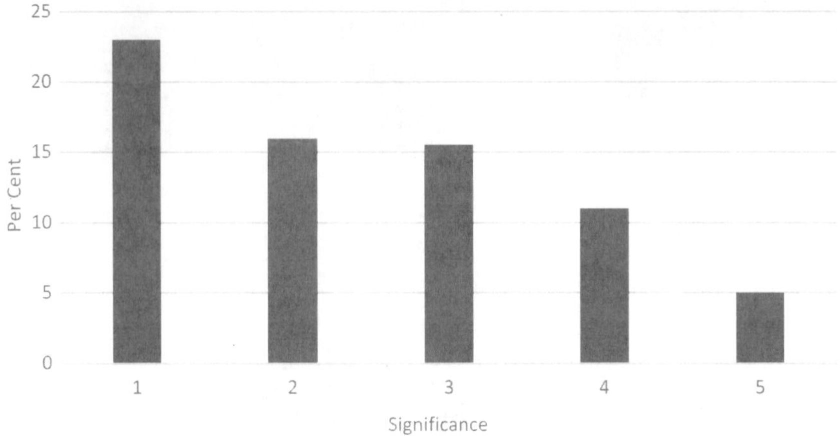

Figure 5.15 Affordable Housing

Respondents were asked if an absence of affordable housing, which was driving pressure for house building, was a significant driver for climate change in Cornwall. Figure 5.15 shows that 23 per cent said that an absence of affordable housing was a very significant driver for climate change in Cornwall. Sixteen per cent said that an absence of affordable housing was a significant driver for climate change in Cornwall; 15.5 per cent said it was of moderate significance; 11 per cent indicated little significance; and 5 per cent indicated no significance at all.

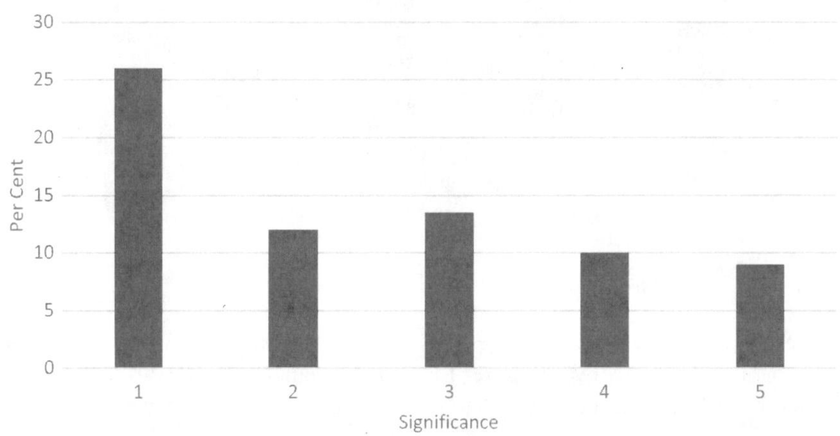

Figure 5.16 Flood Plains

Respondents were asked to indicate if they felt that building structures located on flood plains was a driver for global warming. Figure 5.16 shows that 26 per cent said that building structures located on flood plains was a very significant driver for climate change; 12 per cent said it was a significant driver for global warming; 13.5 per cent said that it was a moderately significant driver for climate change; 10 per cent stated that it was of little significance as a driver for global warming; and 9 per cent said that it was of no significance at all as a driver for climate change.

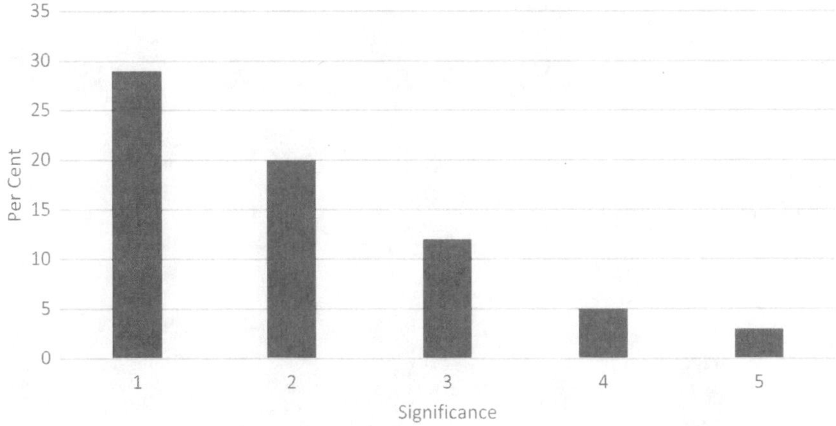

Figure 5.17 Communities Not Self-Sustaining

Members of town and parish councils were asked to indicate whether they felt that the dispersed settlements of Cornwall no longer support self-sustaining communities and lead to increased travel. Figure 5.17 shows that 31 per cent indicated that non-sustainable communities was a very significant driver to climate change; 21 per cent said that non-sustaining communities was a significant driver for climate change; 12.5 per cent said that non-sustaining communities was moderately significant as a driver to climate change; 4 per cent said that non-sustaining communities was of little significance as a driver to climate change; 2 per cent said that non-sustaining communities was of no significance as a driver to climate change.

Members of town and parish councils were asked if they felt that ambitions towards car ownership were a significant driver of climate change. Figure 5.18 shows that 29 per cent said that it was a very significant driver to climate change; 20 per cent said that it was a significant driver to climate change; 12 per cent said that it was moderately significant as a driver to climate change; 5 per cent indicated that it was of little significance as a driver to climate change; and 3 per cent said that it was of no significance as a driver to climate change.

Members of town and parish councils were asked if they felt that an insufficient appraisal and analysis of greenhouse gas emissions was a significant driver for climate change. Figure 5.19 shows that 20 per cent said the insufficient appraisal of greenhouse gases was a very significant driver for climate change; 18.5 per cent

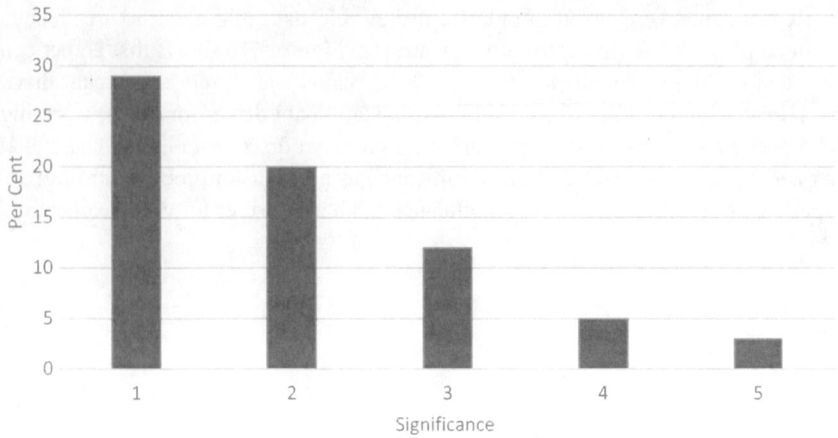

Figure 5.18 Car Ownership

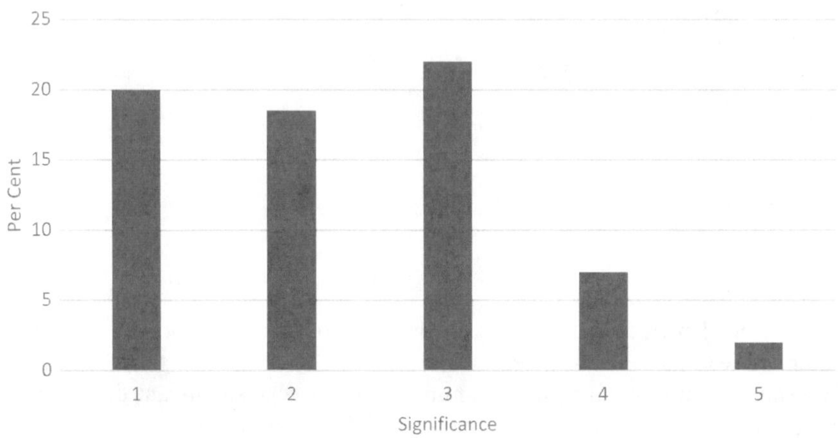

Figure 5.19 Insufficient Appraisal of Greenhouse Gases

indicated that it was a significant driver to climate change; 22 per cent stated that it was of moderate significance as a driver to climate change; 7 per cent said that the insufficient appraisal of greenhouse gases was of little significance as a driver for climate change; and 2 per cent said that the insufficient appraisal and analysis of greenhouse gases was of no significance at all as a driver for climate change.

Overall there are numerous issues to draw from these results. Identification of housing as a concern recognises Cornwall's inefficient housing stock that can contribute to local emissions. This is an issue that is seen more broadly on a national level, with houses contributing significantly to greenhouse gas emissions (CCC 2020; Coley et al. 2012). Within the context of the survey housing was also framed as a social necessity with the lack of social housing seen as a driver for unsustainable development. This is of course a consistent theme when it comes to addressing basic human needs within environmental limits. Similarly, concerns are evident with regard to development on flood plains where the immediate planning pressures conflict with the long-term risks of climate adaptation (Environment Agency 2020; Sayers et al. 2015).

There is evidence within the survey of the connection of climate pressures with Cornwall's distinctive settlement patterns. The dispersed nature of conurbations and non-sustainable settlements is linked with higher demand for car ownership. This reinforces the recognition that rural residents can face systemic barriers to their ability to reduce carbon emissions (Department for Transport 2021). Extending this more broadly the car dependence highlights the spatial inequalities of being able to embed sustainable development in rural communities. The final issue of greenhouse gas measurement reflects the inefficient capacity in carbon measurement and accounting at the time of data collection (Bulkeley & Kern 2006).

Taken as a whole, councillor perspective on the drivers of global warming helps us understand how global risks are translated into relevant aspects of Cornish life. Instead of treating global warming as abstract we can identify tangible place-based pressures and connect this to the broader perceptions of sustainable development. By so doing we connect with broader observations that suggest that local actors interpret global warming through visible infrastructure and experience, which can be highly contextual and variable (Adger et al. 2009; Millar et al. 2025; Prosser & Whitmarsh 2022; Tenbrink & Willcock 2023; Whitmarsh 2005). And as we see increasingly mapping of opinions towards climate change (Gillreath-Brown et al. 2025) and associated risks (Howarth et al. 2025), as well as the ability to analyse ever-greater amounts of data on the subject, it has never been more important to consider context and framework.

Extending the reflexive modernity findings highlight that global risks are both locally interpreted and contested at the parish level. We see that councillors do not passively accept climate discourse but reframe abstract risks in terms of local imperatives. I will not elaborate on this important point at this stage; we will revisit this in Chapter 7.

Global and Local Connection

Whilst complex, an individual's willingness to engage with broader social and political issues and structures depends to certain degree on the importance and relevance that an individual places on perceived risk. As we have seen, the processes through which risk is transmitted from the global to the local sphere are convoluted, uncertain, non-linear, and laden with cultural norms. Directly addressing the global-local connection, town and parish councillors were asked to indicate whether they felt certain issues were firstly, important to the local area and secondly, important on a global basis.

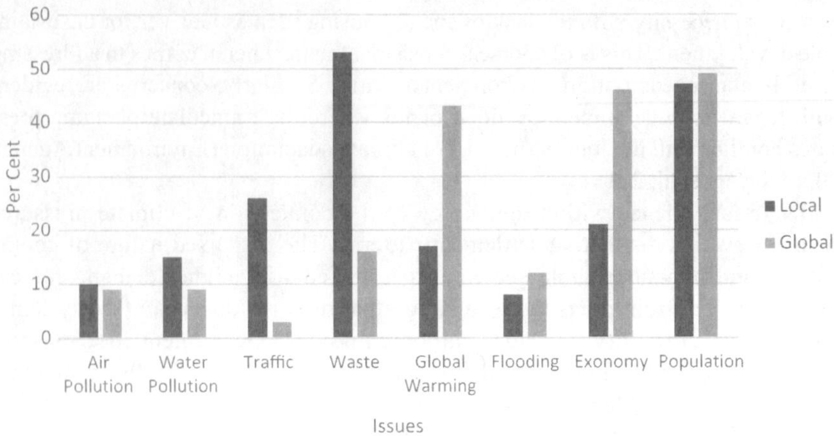

Figure 5.20 Global and Local Connections

As such, respondents were presented with a number of risks that they were asked to identify with at the local and the global level of analysis. Respondents were asked to indicate if these issues were important to the local area and the global area. What is evident from Figure 5.20 is that there are some clear disparities between risk assignment priorities.

Air pollution was considered to be of low importance at both the global and local level; 10 per cent of respondents said that air pollution was important to the local area; and 9 per cent indicated that air pollution was important on a global basis. Water pollution was also considered to be of importance to a small number of respondents at both the global and the local level; 15 per cent said that water pollution was important to the local area; and 9 per cent indicated that water pollution was important on a global basis. Surprisingly concern for flooding was low at both the global and the local level; 8 per cent said that flooding was a concern in the local area; and 12 per cent said that it was of concern on a global basis. Traffic presented a significant disparity between the global and the local area; 26 per cent of respondents stated that traffic was an important issue for the local area; only 3 per cent of participants said that traffic was important on a global basis, 53 per cent of the members of town and parish councils said that waste was a concern for the local area; and 16 per cent said that waste was important for the global area.

Global warming saw a significant disparity between the global and the local area; 17 per cent of respondents said that global warming was important for the local area. The majority of respondents, at 43 per cent, indicated that global warming was important on a global basis. The economy was seen to be more of a concern for the global area, at 46 per cent, than the local area at 21 per cent. Population was seen as a significant concern at both the global and the local area; 47 per cent of respondents said that population was important to the local area; and 49 per cent indicated that population was important on a global basis.

Overall, these results present a disparity between the way that parish councillors prioritise risk locally and globally. Concerns such as traffic were strong at the local level but viewed as insignificant at the global level. On the opposite end of the spectrum, issues such as global warming and the economy were seen as pressing in the global level but less so locally. To a certain degree we are witnessing a progressive reinforcement of some key issues as we have moved through the survey results. Perhaps most pertinent is the recognition that public and community interpretations of risk are framed through experience and issues of immediate visibility. With that said we recognise that these issues are, as we have already explored, mediated through science/lay interpretations, media imagery, and personal experience.

A final observation to make is the variability of responses within the survey that we see in relation to the same issues. With regard to global warming this result presents an interesting insight into how the framing of the questions impacts participant responses. In Figure 5.13 we see that 68 per cent of councillors indicated that global warming was important to Cornwall. This suggests that councillors recognise global warming as a major risk, but when presented with other risks and asked to identify in a global-local context, there is a relative reframing of risk perception. This also occurs for other issues, for example, in climate drivers we saw that 36 per cent saw building on flood plains as a very significant climate risk but in a global-local context this was reduced to 8 per cent. Again risk perception has shifted when framed in different contexts. Other instances within the results include waste and economy in a sustainable development pillars context.

Conclusion

This chapter has progressively explored a range of issues that further our understanding of the relationship between sustainable development and a reflexive modernity. The survey questions and the results are active components in relating sustainable development perception to identified risks and as a result, reflexivity indicators. The survey has provided invaluable insights into this relationship and for the first time presents research data that explores multiple interconnecting variables as understood at the local government level.

The findings suggest that there seems to be significant awareness of the term sustainable development, but that associated mechanisms and practical applications are inconsistent and uncertain. We can see at the general level that councillors associate sustainable development with environmental concerns, though when related to Cornwall economic factors are prioritised. This could point to a certain degree of cognitive dissonance and ambiguity as to how sustainable development is both conceptualised and operationalised at the local level.

Similar responses to risk, and in this instance particularly global warming, also highlight tensions and conflicting values. As we can see most of the respondents acknowledge that humanity is having an impact on global warming and that it is an important issue for Cornwall. But a substantial proportion of survey respondents seemed to doubt the relevance of issues associated with global warming to their communities, or were uncertain how to respond at all. In light of this it could be suggested that whilst global risks are acknowledged, the process of translating those risks into local governance action is uncertain and uneven. This is reinforced by a lack of awareness of the formal sustainable development mechanisms that were presented, and even fewer respondents felt that they are actively involved with these mechanisms. What then do these findings tell us about the relationship between sustainable development and a reflexive modernity?

Initially it is reasonable to say that there is a partial and incomplete presence of a reflexive modernity within the context of town and parish councils. Some elements of the reflexivity indicators that we have identified seem to be clearly present. These include global and local awareness, ethical reasoning, and, to a certain degree, reflexive decision-making. But with that in mind there appears to be a distinct lack of institutional connection to those structures that may embed sustainable development in the community. Within the context of a risk society this presents us with both the potential and the limitations for councils to contribute to embedding sustainable development in their local communities and prepares the ground to explore more detailed qualitative discourses in the subsequent chapters.

Bibliography

Adger, W. N., Dessai, S., Goulden, M., Hulme, M., Lorenzoni, I., Nelson, D. R., … & Wreford, A. (2009). Are there social limits to adaptation to climate change? *Climatic Change*, *93*(3), 335–354.

Beck, U. (2016). *The metamorphosis of the world*. Polity Press.

Beck, U., Bonss, W., & Lau, C. (2003). The theory of reflexive modernization: Problematic, hypotheses and research programme. *Theory, Culture & Society*, *20*(2), 1–33.

Bulkeley, H., & Kern, K. (2006). Local government and the governing of climate change in Germany and the UK. *Urban Studies*, *43*(12), 2237–2259.

Coley, D., Kershaw, T., & Eames, M. (2012). A comparison of structural and behavioural adaptations to future proofing buildings against higher temperatures. *Building and Environment*, *55*, 159–166.

Committee on Climate Change (CCC). (2020). UK housing: Fit for the future? Committee on climate change. https://www.theccc.org.uk/publication/uk-housing-fit-for-the-future

Department for Transport. (2021). Transport and environment statistics 2021 annual report. GOV.UK. https://www.gov.uk/government/statistics/transport-and-environment-statistics-2021

Environment Agency. (2020). Flood risk assessment: Climate change allowances. GOV.UK. https://www.gov.uk/guidance/flood-risk-assessments-climate-change-allowances

Fischer, F. (2000). *Citizens, experts, and the environment: The politics of local knowledge*. Duke University Press.

Gillreath-Brown, A., Goddard, E., Jefferson, M., Richards, E., Carman, J., Rosenthal, S., Leiserowitz, A., & Marlon, J. (2025). UK climate opinion maps 2024. https://climatecommunication.yale.edu/visualizations-data/ycom-uk/

Howarth, C., Zamzow, H., McLoughlin, N., & Shreedhar, G. (2025). Gaps in public perceptions of extreme heat highlight UK's lack of preparedness. *Environmental Science & Policy*, *170*, 104101.

Hulme, M. (2009). *Why we disagree about climate change: Understanding controversy, inaction and opportunity*. Cambridge University Press.

Hulme, M. (2023). *Climate change isn't everything: Liberating climate politics from alarmism*. John Wiley & Sons.

Millar, C., Mulholland, C., Zanin, B., Whitmarsh, L., Gibson, R., Meyer, J., & Demski, C. (2025). *A review of effective public engagement on climate and implications for Scotland.* https://era.ed.ac.uk/items/d4162def-bb54-446f-9991-66c5addd170d

Prosser, A., & Whitmarsh, L. (2022). Net Zero: A review of public attitudes & behaviours. The University of Bath for Our 2050 World. https://researchportal.bath.ac.uk/en/publications/net-zero-a-review-of-public-attitudes-amp-behaviours/

Purvis, B., Mao, Y., & Robinson, D. (2019). Three pillars of sustainability: In search of conceptual origins. *Sustainability Science*, *14*(3), 681–695.

Sayers, P. B., Horritt, M. S., Penning-Rowsell, E., & McKenzie, A. (2015). *Climate change risk assessment 2017*. Committee on Climate Change.

Stamm, K., Clark, F., & Eblacas, P. (2000). Mass communication and public understanding of environmental problems: The case of global warming. *Public Understanding of Science*, *9*(3), 219–238.

Telleria, J., & Garcia-Arias, J. (2022). The fantasmatic narrative of 'sustainable development'. A political analysis of the 2030 Global Development Agenda. *Environment and Planning C: Politics and Space*, *40*(1), 241–259.

Tenbrink, T., & Willcock, S. (2023). Place attachment and perception of climate change as a threat in rural and urban areas. *PLoS One*, *18*(9), e0290354.

Whitmarsh, L. (2005). A study of public understanding of and response to climate change in the South of England (Doctoral dissertation, University of Bath).

Whitmarsh, L. (2009). What's in a name? Commonalities and differences in public understanding of "climate change" and "global warming". *Public Understanding of Science*, *18*(4), 401–420.

Wilson, K. (2000). Communicating climate change through the media: Predictions, politics and perceptions of risk. In S. Allan, B. Adam, & C. Carter (Eds.), *Environmental risks and the media* (pp. 201–217). Routledge.

6 Internalising Sustainable Development

At the outset, it is recognised that throughout the corpus of interview transcripts, there is uncertainty over what the concept of sustainable development entails and how it is being applied both nationally and locally. This illustrates a particular disconnect between how the concept relates to policy and how the economic, social, and environmental dimensions of sustainable development interlink. What follows is presented in the form of a fluid narrative designed to build a progressive story of the themes that emerged. It is recognised here that inevitably many important issues discussed by the interviewees lie beyond the scope of this chapter, and indeed this sentiment is true of all the characters that present qualitative insights. The omission of these issues reflects the need to provide a coherent focus on the key goals of this book. The narrative presented here therefore is recognised as one critical strand within a broader and more complex picture. As discussed in Chapter 4, the interview schedule presented a loose outline of key information that I wanted to discuss in the interviews. Very few interviewees discussed only these issues, with interviewees' priorities, passions, insights, and observations being juxtaposed in the narratives they presented.

In the Beginning

Interviews began to articulate their understanding of sustainable development within the context of some easily identifiable categories. For example, environmental conservation inevitably arose as a key issue. This interviewee identified a key geographical theme that arose repeatedly throughout the interviews. This is related to Cornwall's extensive coastline and maritime focus.

> We're also Heritage Coasts, we've got so many categories which is protecting the area, we're concerned that the roads are not wide enough for some of the development in some of the areas they want to put a pile of houses.

Here the interviewee discusses both the issue of heritage coast as well as development, infrastructure, and concerns over housebuilding. We will reengage with narrative of ocean and coastline more substantially in Chapter 8. At this point we recognise that the interconnections between lived experience, heritage, and

DOI: 10.4324/9780429297199-6

landscape echo broader observations which demonstrate how local identity is deeply tied to memory and place, as well as the relationship between natural and altered landscapes (Wheeler 2014). The issue of overpopulation, especially in relation to seasonal tourism spikes and the impact that this had on interviewees, was visible through the number of cars that cause congestion and safety-related issues. As this interviewee states:

> The streets are cluttered with cars, how emergency service will get through some of the streets I don't know, there's no sense to the parking… we just had to have the streets marked out here with double-yellow lines and parking bay because next door… I could show you photographs of my entrance, and it's a joint entrance, all blocked off with parked vehicles right the way across, and I have to be asked if I can be let out.

This sense of restriction in a space that is often presented as an area of wilderness and freedom is a stark contrast to the realities that local residents experience. This then forms an initial insight into an underlying tension within the sustainable development narrative and is positioned within established observations of the perceived and lived experiences of rural inhabitants (Shucksmith 2018). But this tension was consistently tempered with expressions of awe and wonder for the natural landscape that permeated the aforementioned narrative. The following excerpt personified this perspective:

> We've got a beautiful coastline, very, very rugged but amazing, and if you walk on the coastal path and you look down and you see the sun on the sea rippling, it's like looking down on a basket of jewels.

Again the coastline emerges as a key factor in interviewees' understanding of the environmental and the sustainable development discourse. The description of an almost ethereal image of the ocean as a basket of jewels resonates with Macnaghten and Urry's (1998) observations that such moments of awe embody a cultural ambivalence of nature in late modernity. Ambivalence in the sense that nature is at once sublime and threatened, free but also constrained by risk. Further engagement with sustainable development, this narrative also represented one of the underlying principles of sustainable development in relation to intergenerational equity. As this councillor remarks, *The beaches around here are just stunning, and it's crucial to keep them that way for future generations.* Whether interviewees were aware of this key principle or not, it emerged as a consistent and overt discourse throughout the interview material. This initially demonstrated a concern for future scenarios but not necessarily sustainable development itself. Many more issues emerged and these will be explored later in this chapter and throughout the book.

What is more, these narratives moved beyond the passive identification of the landscape to the active engagement with the natural environment. This interviewee points to surfing as an activity that has raised their awareness of the ocean and the natural world. *There are so many people, locals and tourists using our waters, obviously surfing is part of the Cornish identity and I think this direct connection to the sea actually increases understanding of important issues, climate change, pollution, sewage, amongst other things.* Elsewhere, I have commented on the role of engagement with nature and especially within the context of ocean surfing (Borne 2018), and more broadly, research has focused on how outdoor recreation is impacting environmental attitudes and behaviours (Alcock et al. 2020; Martin et al. 2020). Whether on land or in the ocean, there is increasing evidence that connection to nature has a significant impact on an individual's relationship with the environment and their action towards it (DeVille et al. 2021).

In tandem with the aforementioned issues, economic necessity was also clearly prevalent within the sustainable development discourse and this touched on a number of relevant issues, locally, nationally, and internationally. In a single sentence the following interviewee succinctly touches on a number of issues. *There's unemployment, I mean, a lot of people I've been talking to in Cornwall say there's a lot of isolation and deprivation and that word about poverty, whatever that means nowadays.* Isolation, deprivation, poverty, and unemployment ran through narratives of sustainable development often intertwined with the discussions on nature and the environment. These structural issues are well documented and were discussed in Chapter 3 where rural deprivation and exclusion in the UK are increasingly represented as a multi-dimensional issue (Shucksmith & Brown 2016).

The following interviewee discusses the issues relating to housing and bank lending. *I think that times are very hard at the moment for people, particularly those buying their properties, yes they've got deprivation because banks have allowed them to be overstretched.* Other early issues that arose, which considering the interview group was not surprising, related to the role of funding and social services and the relationship between local and national government. All of these issues will be elaborated on in due course. And in various ways these observations respond with the contextual discussion in Chapter 3 as well as broader literature that relate to debates on austerity and centralisation in rural governance (Copus & Clark 2020; Lowndes & Gardner 2016; Scott et al. 2019; Tomaney et al. 2010).

Another theme that emerged in association with sustainable development discourse is community resilience. Community, from the outset was a key issue within the research and has already formed a consistent theme in this book, indeed Chapter 8 elaborates on community dynamics in detail. Also, the survey data presented in Chapter 5 has already laid the foundation for a broad understanding of community and the discussion presented here builds on those observations as well as being a stepping stone for the subsequent discussion in Chapter 8. Interviews demonstrated themes that are consistent in many discussions on the nature of community. These include initial definitions of community itself – what does community mean within any particular context? Changes in community dynamics at multiple levels. Activities within communities that change over time that are important for a sense of community cohesion were emphasised, in subsequent discussion we will explore this issue in relation to the fishing industry in Cornwall. Partnership between community members and through partnership with other organisations and bodies was also present very early on in the interview process. These observations reinforced the findings identified in relation to partnership within the survey. There was a consistent theme relating to resource sharing and collaboration in local government and service delivery.

Against this backdrop, community resilience can be seen to represent both an outcome of, and a driver of sustainable development (Berkes & Ross 2013; Garcia & Vale 2017; Magis 2010). Also we can begin to recognise narratives of community in relation to sustainable development within the context of assemblages we identified in Chapter 3. That is where the definitions and practices of community are recognised as being fluid, contested, and adaptive. Moreover, the emphasis on partnerships and resource sharing positions these discourses within a reflexive governance framework that prioritise collaboration under conditions of uncertainty.

Permeating all of the aforementioned issues, interviewees highlighted the importance of a sense of community and belonging. As members of the parish councils it is perhaps an obvious finding that the interviewees as a group are actively engaged, on a voluntary basis, in their local communities and represent a conduit from and to the country's political apparatus. But the sense of community as it was presented within the context of sustainable development discourses, presented in different ways. Early dynamics emerged that allowed for definite themes to emerge. Community resilience is used here as an overarching theme as within this theme multiple issues emerge in relation to sustainable development.

> I think there's quite a strong sense of community yes. We've got a mix here of the old village people, when I say old, people who are part of generations that lived here, and then a lot of incomers such as myself.

This interviewee identifies a strong sense of community in their parish area. There is an interesting reference within this excerpt to both long-standing members of the community and the new members of the community and those that are not indigenous to Cornwall. This tension represents a significant demographic dynamic that is not unique within Cornwall, but is prevalent in rural and coastal areas. These dynamics also relate directly to narratives of community resilience, continuing to reinforce the assertion that sustainable development is lived through the everyday interplay between identity and place.

In sum, the previous discussion provides an initial snapshot of what is in reality a hugely diverse and often contradictory narrative in relation to sustainable development. It is worth reminding ourselves at this point that perceptions of sustainable development were being sought in a period that predates the sustainable development goals (SDGs), that predates the privileged conceptual status that sustainable development now occupies. The research period also predated the proliferation of frameworks and evaluative tools that have emerged since the establishment of the SDGs in 2015. This is not to say that we are now in an era of acceptance, conformity, and universal implementation, far from it, we are however in an era that, at the very least, recognises the importance concept, and across sectors and nations, there is a need to engage with the tensions incumbent within it either directly or indirectly. With that understanding in mind we now move to draw out, what was a very central theme that consistently presented itself throughout the interviews.

Scepticism

Scepticism emerged as a key issue within the interviews representing a lack of salience with the concept of sustainable development. This broadly occupied two camps. One was with the terminology itself and the second was with the practical implementation of the concept of sustainable development.

Scepticism was often presented within the context of the term being overly abstract.

For example, one interviewee indicated that

> I don't like the phrase I'll be quite honest to start off with. I think it is, it's a nice all-encompassing phrase but as far as the community is concerned it means people can live in that community and not actually travel far to get what they want.

Whilst this interviewee acknowledged that it was abstract, they also related the concept to community and the way that community was understood. And this resonated with the broader corpus of data with a general lack of understanding of the term but an underlying narrative that related to community. The following interviewee reinforces that perspective. *It's a funny word isn't it, sustain, I would have thought something more to do with the community, a community word, you know a word that sounds more like enhancing the community as opposed to sustainable blah, blah, blah.*

Community then provided a foundation from which interviewees could build their understanding of sustainable development. Given the context of parish councils with a strong emphasis on community this is not a surprising finding and adds weight to the need to further engage the local to address global issues. But within this context multiple and diverse issues evolve. For example, the relationship with sustainable development, community, and business emerged consistently throughout the data. The following interviewee is representative of this narrative.

> If you create a business, that business has to pay and it's not draining off from the bank, I think with a community, it's not going to rely on government 100%, it's got to rely on itself within the community and that's where I see sustainable development.

This councillor addresses sustainability within the context of ensuring the business is able to sustain itself into the future but also be viable without the support of government funding or subsidies to support its operation.

This observation uncovers the relationship between local government and the business community. It raises the question of what type of relationship exists between different sectors, how funding is allocated and what product, services, and broader issues are seen as important in supporting community cohesion. This tension presents itself throughout the research and continues to represent an ever-changing funding landscape.

We can reengage with this point with the broader political landscape as well as the previous discussions presented in Chapter 3. This is especially relevant as the transition of power between a conservative and labour government represents not only a change in policy and priority but also a philosophical position. The new labour government within its first week in power has engaged with the political philosophy of devolution, though I proviso that statement with observations from Wills (2016) that illustrate the political vagaries of devolution dynamics. Perhaps we witness this engagement with devolution through the removal of the term 'levelling up' and replacing it with local government. Following the Labour Party's victory in the 2024 General Election, the Department for Levelling Up, Housing and Communities was renamed the Ministry of Housing, Communities and Local Government. Moreover, a raft of immediate policy aspirations were introduced that included significant alterations for the National Policy Planning Framework with its emphasis on sustainable development. This agenda is ongoing but will have practical implications that will affect the way that sustainable development is delivered and perceived. At the forefront of this shift in narrative is the importance placed on 'getting Britain building again' to address the housing shortage.

This is a fluctuating policy objective that endures from the time of data collection to contemporary policy objectives. During the period of data collection housing policy was largely shaped by top-down national and regional strategies with a strong emphasis on the use of brownfield sites.

At the parish council level there was a clear recognition of the importance of housing, in the role of community and sustainable development. In that context parish councils operated as consultees with limited formal power to shape outcomes. Today, and as we have seen in Chapter 3, housing policy is guided by National Planning Policy Framework and reinforced through tools like neighbourhood planning and biodiversity net gain. Within this dynamic parish councils move away from simply being consulted and towards new housing aligns with issues such as climate goals, ecological protection, and community priorities. In the context of the changing policy landscape, sustainable development and housing were connected in multiple ways within the interview data. The following interviewee begins to make these connections

> Sustainable development really is looking at problems involved. One of the things that we were hit with is affordable homes yeah and they're obviously going to have to be green affordable homes, yes so look at your village because we want to find land to build affordable.

This excerpt draws a number of issues to the fore. Sustainable development here is seen as a response to problems that exist within the parish area. This makes the connection between the global and national narratives that are abstracted and as we have already seen often viewed with scepticism. This then presents an avenue for communicating the relevance of the concept, but not as a concept in itself but as tethered to the realities of local communities and individuals' lives. In developing this narrative within the context of affordable housing the interviewee describes the mechanism for achieving this in conversation with the district council. The interviewee expressed a desire to work constructively with the district council to achieve this goal by identifying plots of land for potential construction:

> So we said ok, that's fine, that sounds alright we'll do that and if we can help, contribute in any way, so they sent out a form with areas where they'd looked at land without help, there's a plot there, a plot there, a plot there, and I said, one of my questions was, I said who wants the affordable houses?

As is evident from the above excerpt, whilst there was a willingness to achieve the goal and work collaboratively, there was also a fundamental questioning of the need and demand for affordable housing.

> Well we have a waiting list so you know, we looked at the waiting list and I think there was 3 people on the waiting list possible wanting affordable housing you see. The government remit is to build so many affordable houses for each area, the statement was you'll need 5 affordable houses building, to be built so there we are, we looked at the land. The land got turned out, failure in one or two different instances where access to it wasn't right, highways didn't want a road, opening on to a road there and bits and bobs so it all got put on the back shelf, on the back burner, that was it, so we have no affordable housing.

It's clear from the interviewee's statement that there is frustration with the process of identifying the relevant building plot and a disjuncture between the aspiration for affordable housing and the reality of the implementation from an infrastructure and planning perspective. The interviewee expanded on this issue by highlighting the infrastructure needs within Cornwall, accessibility to relevant land, and the restrictions that inevitably occur within a rural environment. The narrative of sustainable development was also expanded to include the actual construction of the affordable housing.

> This affordable housing would have had proper green roofs and things, and proper insulation and all the different things and I would have wanted to have something with some social connection, like you've got a shop and not so isolated.

Here there is mention of green roofs and insulation which relates to both the aesthetics of the housing as well as the energy efficiency of the buildings themselves. So the narrative in connection to sustainable development is represented by this interviewee in a complex and contradictory way. There is a need to communicate with the local authority to create the needed social housing. But there is also recognition of the barriers and constraints to making this happen, from general infrastructure problems through to the specifics of rural community which present their own challenges. Furthermore, there is also recognition of the social connection, the creation of community, and the need to make connections between different areas. The interviewee cautions that without creating an enabling environment for these disparate areas to coexist, progress will be limited.

> We're finishing up with villages that have got nothing, nothing and you now need transport to get to anything, whereas if it had been collective and compacted it would have been self-running in effect, so that's the way I look at that bit.

Inevitably, the narrative returns to the scattered geographical nature of settlements within Cornwall and of course this is representative of rural communities across the country and indeed across the world. This also represents the consistent tension between narratives that present themselves in rural and urban environments. Whilst the SDGs recognise the interconnected nature of creating solutions to sustainable development-related issue, there is nevertheless often a reductionist representation of the issues that are inherent within the rural and urban areas. For example, whilst SDG 11 identifies sustainable cities and communities there is not a specific SDG that identifies rurality. The reference is instead to *making cities and human settlements more inclusive, safe, resilient and sustainable*. As previously mentioned, the interconnected nature of the goals means that many goals address rurality. Table 6.1 briefly demonstrates this.

And within this context of rurality and the issues that rural areas are exposed to other interviewees continues to explore sustainable development within the context of 'the future'. *I suppose it simply means it will stand the test of time in one of many ways. It doesn't necessarily have to be a building that will stop up a long time. It can be a service that is sustainable in the true sense of the word.* Here the interviewee is referring to the sustainable development phenomenon standing the test of time. This is referred to not only in the context of the physicality of objects such as buildings and infrastructure, but also a service that is sustainable in the 'truest sense' of the word. This comment illustrates a consistent narrative, not only within the interviews but more broadly within society about the common sense understanding of sustainability, and then that which relates to sustainable development. The following interviewee echoes this into the future narrative:

> Well sustainable development is obviously put something in place that will last and be of benefit to the community, that could be little business start up units or something and then that there, businesses can stay there be sustainable and provide jobs.

In this excerpt again sustainable development is represented as being something that will last, and as in the previous example it is related to the benefit of the community. Here there is a direct connection between business and the community resonating with previous comments on the relationship between business, community, and the ability of communities to be self-sufficient. These reflections resonate with

Table 6.1 SDGs Relevance to Rural Areas

SDG	*Description*	*Relevance to Rural Areas*
Goal 1: No Poverty	Eradicate extreme poverty and reduce the proportion of people living in poverty.	Rural areas often have higher poverty rates than urban areas.
Goal 2: Zero Hunger	End hunger, achieve food security, improve nutrition, and promote sustainable agriculture.	Rural areas are typically agricultural hubs and are central to achieving food security.
Goal 3: Good Health and Well-being	Ensure healthy lives and promote well-being for all at all ages.	Access to healthcare services in rural areas is essential for this goal.
Goal 4: Quality Education	Ensure inclusive and equitable quality education and promote lifelong learning opportunities.	Improving education in rural areas is critical for their development and community resilience.
Goal 5: Gender Equality	Achieve gender equality and empower all women and girls.	Gender disparities can be more pronounced in rural areas, making this goal significant.
Goal 6: Clean Water and Sanitation	Ensure availability and sustainable management of water and sanitation for all.	Access to clean water and sanitation is often a major issue in rural areas.
Goal 7: Affordable and Clean Energy	Ensure access to affordable, reliable, sustainable, and modern energy for all.	Renewable energy generation, energy security
Goal 8: Decent Work and Economic Growth	Promote sustained, inclusive, and sustainable economic growth, full and productive employment, and decent work for all.	Creating economic opportunities in rural areas is vital for their development.
Goal 9: Industry, Innovation, and Infrastructure	Build resilient infrastructure, promote inclusive and sustainable industrialisation, and foster innovation.	Improving infrastructure in rural areas can enhance connectivity and economic opportunities.
Goal 10: Reduced Inequality	Reduce inequality within and amongst countries.	Addressing disparities between rural and urban areas is part of this goal.
Goal 11: Sustainable Cities and Communities	Make cities and human settlements inclusive, safe, resilient, and sustainable.	Whilst focusing on urban areas, it includes aspects like regional planning that benefit rural areas.
Goal 13: Climate Action	Take urgent action to combat climate change and its impacts.	Rural areas often face significant climate-related challenges that need to be addressed.
Goal 14: Life Below Water	Conserve and sustainably use the oceans, seas, and marine resources.	Relevant to rural coastal and fishing communities.
Goal 15: Life on Land	Protect, restore, and promote sustainable use of terrestrial ecosystems, manage forests, combat desertification, and halt biodiversity loss.	Many rural communities depend on land and natural resources for their livelihoods.

(*Continued*)

Table 6.1 (Continued)

SDG	*Description*	*Relevance to Rural Areas*
Goal 16: Peace, Justice, and Strong Institutions	Promote peaceful and inclusive societies, provide access to justice for all, and build effective, accountable institutions.	Strengthening institutions in rural areas can enhance governance and development outcomes.
Goal 17: Partnerships for the Goals	Strengthen the means of implementation and revitalise the global partnership for sustainable development.	Partnerships can support rural development through resource mobilisation and capacity building.

a broader common sense understanding of sustainability as endurance over time, but the emphasis here is on the benefits of the community instead of abstract global targets. Identifying this within a reflexivity rubric framing sustainable development in terms of self-sufficiency illustrates a pragmatic reflexivity grounding global discourse in everyday practice.

A number of the interviewees expressed sustainable development as previously represented highlighting some key and expected areas. This included social, environmental, and economic issues as they related to Cornwall and to those that were more global in nature. We have already seen in previous chapters that the global and local connection was prominent within the sustainable development discourse and this will be revisited on a number of occasions within this book. Some interviewees also presented discourse of sustainable development that were more in depth, expressing multiple and interconnected issues within the single interview. The following interview from a single interviewee will be examined in more detail. The interview is with a parish councillor in a large parish area in Cornwall. There are a number of issues to be explored within the context of this single interview. The following example, drawn from a councillor in a large Cornish parish, illustrates how sustainable development is embedded within historical, social, and economic contexts that stretch beyond immediate community concerns. Initially the interviewee connects sustainable development to a historical narrative within Cornwall.

> For sustainable development, we're really harking back to 1910, villages were only sustainable because employment was in the village, this situation I can't really ever see occurring again.

There is an immediate connection to a pre-war–pre-industrial Cornwall. The narrative of employability and the localised nature of working in a rural environment are highlighted. Cornwall's industrial base as with many other areas was and is innately connected to its natural resources. Table 6.2 highlights the predominant forms of employment in 1910.

There was then clearly some local resource-based employment, especially fishing, mining, and agriculture, that requires local-specific employment. Also evident from the table are emerging geographically connected industries such as tourism transport and shipping that sets the trajectory for future globalised industry. Recent

Table 6.2 Predominant Employment

Employment Sector	*Description*
Mining	*Tin Mining*: Significant but declining industry, recent investment. *Copper Mining*: Also in decline but provided jobs.
Agriculture	Farming, dairying, and livestock management were major sources of employment.
Fishing	Coastal communities engaged in pilchard fishing and other types of fishing.
China clay (Kaolin) extraction	Major industry, especially in the St Austell area, for ceramics production.
Quarrying	Extraction of granite and slate for construction and other uses.
Manufacturing and craftsmanship	Small-scale production of tinplate, metal goods, weaving, pottery, and blacksmithing.
Tourism	Early stages of tourism industry with jobs in hospitality, guiding, and visitor services.
Service industry	Employment in domestic service, retail, and other support services, especially in towns.
Transport and shipping	Jobs in ports, railways, and related to the movement of goods and people.

initiatives and investment by the UK government are seeking to reinvigorate Tin mining in Cornwall by reopening the South Crofty mine which closed in 1998. An investment of nearly £29 million to Cornish Metals who own the mine is a response to increased demand for Tin in electronic components. Moreover, an additional £4.19 million has been allocated to develop infrastructure around the mine. This has been awarded from the Shared Prosperity Funds Good Growth Programme. Returning to the interviewee and building on this historical connection, there is a continued expansion of their understanding of sustainable development.

> We've all become used to getting what we need from a retailer, and to get to a retailer you need transport, now if transport runs out there's no way of getting to a retailer. So sustainability to be brutally frank won't happen in my opinion, the only way to make it happen is to become self-sufficient by growing all your own vegetables and meat and so on and so forth.

Focusing on retailing the interviewee expresses concern about the ability for a sustainable future to be achieved in Cornwall. This relates directly to transport, which as was identified within the survey data was a predominant issue for respondents. For this interviewee and with previous interviewees the idea of self-sufficiency emerges as a potential solution and criteria for achieving sustainable development. Instead of creating a sustainable business as identified previously, the focus here is on food where the narrative of bioregionalism comes to the fore. This bioregional focus can be drawn out of a number of interviewees and connects to the philosophical and epistemological context of what has been described as a strong interpretation of sustainable development and particularly environmentalism (Neumayer 2012). However, in the midst of this bioregional enthusiasm, the interviewee also expresses connections to broader global processes. In particular the narrative turns to finite resources and the planet's carrying capacity.

> Well you come down to basics, the world's overpopulated and that really is the end of the line, we've reached the stage that unless we get plagues or wars, and wars have always been around, the situation isn't going to right itself, and wars by the very nature work the wrong way round in the end, you reduce your population but at the end of it you increase it dramatically so really it's way beyond us.

Where earlier interviews identified in this chapter emphasised the local consequences of population pressures, this interviewee expanded on global limits raising questions of carrying capacity and planetary limits. For this interviewee overpopulation is a fundamental issue, it is 'down to basics' (Meadows et al. 1972; Ehrlich 1968; O'Neill et al. 2017). And of course, these reflections resonate with global debates. For example, the UN's World Population Prospects 2022 highlights how rapid population growth intensifies pressures on finite resources (UN DESA 2022), whilst the Human Development Report 2023/24 frames these demographic and ecological stresses as fundamentally political, requiring new forms of global cooperation (UNDP 2023).

There is a fundamental reality when addressing multiple issues that relate to sustainable development that is underpinned by a rapidly increasing global population. The inequitable distribution of resources, the inequitable areas of investment growth and prosperity and the continued concentration of wealth in fewer and fewer hands. Interestingly, the interviewee then presents a contradictory narrative around the role of conflict in addressing overpopulation. Whilst somewhat dramatic and inconsistent this reflection does highlight a broader theme that emerged from the interview data. That is an awareness that global challenges are an integral part of the narrative of sustainable development, combined with uncertainty and confusion about how these issues are addressed in practice. The following interviewee extends this uncertainty refocusing the narrative back to the local level as a means of making sense of a global problem:

> But I really don't know, in the meantime all we can do is apply a palliative or a start to local development that doesn't cost the country for ever, which comes back to gardening.

The term palliative presents an interesting insight into the overall perspective. Palliative is a term that is often related to end-of-life care where a patient is made comfortable but the underlying illness and cause of their demise is beyond treatment. Applied then to the sustainable development narrative, this interviewee feels that the current trajectory is beyond a cure. The conversation then moves to the relationship between the rural and the urban environments with a particular focus on energy production and consumption, the policy context of which was discussed in Chapter 3, and again there is an emphasis on local production.

> Cities they can each have a small nuclear power station and harness the wind or the river, I mean some cities could harvest the rivers and that would secure that side of the equation. The transport can all be electric, so they've got less problems in a way. Then they can have community farms around the periphery, instead of having a greenbelt we can have a garden belt, but the greenbelt only came in after the war. I think it's all dramatic basic thinking and I can't honestly see it happening.

There is a prescience to this except that it reveals both optimism and a reluctance to move beyond a palliative perspective. This point therefore we begin to make connections between the time of data collection and the realities of today's world. Firstly, there is an emphasis on the need for renewable energy from localised nuclear power through to wind and hydro energy production. The transition towards electrification of cars is also suggested in the intervening years between the time of data collection and today, there have been mixed trajectories towards these sustainable futures (Sovacool 2016). These trajectories have, as is predominantly the case been subject to the global socio-technical context that has fluctuated within the broader political ideologies (Geels 2011; Grin et al. 2010; Smith et al. 2005). The uptake of electrified vehicles has progressed substantially in the past ten years to a global position where technology has advanced to make electric vehicles viable, and this has kept pace with a public and political willingness for this transition. The conservative government's moratorium on onshore wind power has, to a certain degree stifled the progress of this technology, though advancements in technology and infrastructure have been made.

The moratorium was lifted on onshore wind farms days after the labour government was voted into power representing a significant shift in policy and ideology. These debates over energy transitions connect directly to broader planning reforms, where shifting political ideologies continue to shape the balance between development and sustainability. This resonates strongly with the interviewees' call for a garden belt to accommodate localised food production. Instead of a garden belt, the Prime Minister Sir Kier Starmer's government has proposed a 'grey belt'. This grey belt will include redesignating areas of green belt considered of a lesser quality, this includes areas such as car parks, quarries, and shrub land.

What we draw from this, as was highlighted at the outset of this book is a reengagement with the broader political and ideological landscape as the UK returns to a labour-dominated parliament. Presenting a similar assessment, the following interviewee identifies a particular perspective on sustainable development. Initially, the interviewee articulates their general understanding of sustainable development. *Well I think sustainable development means development that can fit in and maintain what we've got as well as giving the infrastructure if you like and the wherewithal to maintain to the community.* Maintaining the status quo is the emphasis of this in combination with the community and infrastructure. The interviewee then expands on this foundational narrative to include attracting families into the parish areas.

> In our village we need to be able to encourage some families to come and live here so sustainable development would be looking at what we've got within our parish and seeing how we can perhaps allow some affordable family housing.

Sustainable development in this instance relates to family and affordable housing. However, the interviewee cautions that attracting families and the process of social housing development should be consistent with local values and sensitively implemented.

> Now it's got to be done in such a way that it doesn't tear out the wherewithal, so it would, for example if you take some of the local farms and you just build a big estate and now I don't think that's what we're talking about.

The interviewee uses the phrase wherewithal to articulate the local context. There is also an emphasis that simply identifying farmland for development is not an effective or appropriate way to achieve the social housing goal.

> We're talking about perhaps subtle development, sympathetic development, and trying to knit the framework together, if we get some families in let's look at getting some jobs for the families, let's get some transport to help the families, let's pull it together in a joined up way, that's what we need to do, so that's about sustainability I think.

The interview recognises that potentially disparate areas need to be connected in order to effectively address sustainable development within their parish. For the interviewee, this involves 'subtle and sympathetic development'. Employment and families go hand in hand as does infrastructure and transport. This systems approach to local service delivery sits at the core of effectively enhancing areas within the Cornish Parish. We have already explored local perceptions and understanding of the opportunities and barriers to local service delivery in Chapter 4.

The interviewee also explored to what extent the terminology of sustainable development helps catalyse individual and organisational perspectives around creating a more joined-up and effective approach. The interviewee expresses a hopeful aspiration that sustainable development will create a conduit for creating a vehicle for collaboration.

> I think sustainability you know has the ability to draw people together, but unfortunately there's so many buzz words aren't there and I think sometimes people hear it and it just becomes a buzz word, until we find something better I suppose sustainability is the one.

These final remarks represent the tension within the general corpus of interviewee data.

Sustainable development has been framed not so much as an abstract global concept but more in connection to issues of housing, employment, transport, and community cohesion. The emphasis on a subtle and sympathetic development suggests an understanding of sustainable development that is more about integration than it is about expansion. With that said, the concern for buzzwords further points to the tensions between everyday narrative and lived experience and political discourse. And as we reflect on narratives and discussions presented in this chapter, interviewees' insights highlight both the potential for sustainable development to become a unifying narrative and the limits of its relevance at the local level.

Conclusion

This chapter builds on the survey analysis in Chapter 5 to provide a more grounded and experiential understanding of sustainable development. The interviews' narratives emphasise uncertainty and scepticism about sustainable development as an overarching concept. But with that said, the interview simultaneously reveals a deep and often instinctive understanding of the core principles that are associated with sustainable development. This is evident in values such as intergenerational equity, environmental stewardship, and a clear sentiment of the need to protect community cohesion. There is little doubt that environmental issues and themes features heavily throughout the interviews, inevitable references and connections are repeatedly made to Cornwall's coastline, its rurality, and broader issues of nature connection. There was often a general awareness of climate-related risk and many interviewees framed their responses and concerns through localised issues such as flooding, overdevelopment, infrastructure capacity, and transport. These issues were often expressed through emotional and sensory language which underscores the personal and lived connection to these risks. Community resilience, though often not interpreted in those terms, emerged as a critical theme, representing shifting demographics, social fragmentation, and changing economic priorities. As is evident interviewees frequently associated sustainable development to issues of affordable housing, employment, and the viability of local service delivery. Again, these connections were often framed through personal experience, historical memory, and distinct connection to Cornish heritage.

Engaging with the reflexive modern indicators, the interviews reinforced a number of these underlying connections. For example, narratives that represent a displacement of traditional authority were evident across the interviewee data. This was partially visible in the context of critiques of national policy-making and planning processes. We can also see evidence of knowledge pluralism through challenges to expert-led narratives and the preferences for local understanding and context. Reflexive choice-making can be seen running through the interview data as a consistent narrative. Calls for locally managed development, resource use, and transport solutions. Ethical reasoning manifests in different contexts relating to fairness, housing, and intergenerational responsibility. And as previously identified, tensions persist between the aspirational nature of sustainable development and the practical limitations felt on the ground. There was a particular perception of a disconnect between government targets and local realities. However, there was also optimism that sustainable development could become a meaningful framework if connected to the lived experience of local communities.

The interviews point to a situation where sustainable development is being internalised not through mechanisms of abstract policy making, but through the deeply held local dynamics. The concept becomes relevant where it resonates with personal values, community priorities, and local memory and heritage. In light of this, I suggest that what we see is a partial but genuine movement towards a reflexive modernity. But this is a reflexive modernity that is not shaped by formal structures and instead forged through the everyday reflection and practise of those councillors that grapple with the reality of community governance. These observations lead us directly into the next chapter where we turn more directly to questions of risk, uncertainty, and how global warming shapes the way sustainable development is understood.

Bibliography

Alcock, I., White, M. P., Pahl, S., Duarte-Davidson, R., & Fleming, L. E. (2020). Associations between pro-environmental behaviour and neighbourhood nature, nature visit frequency and nature appreciation: Evidence from a nationally representative survey in England. *Environment International*, 136, 2–10.

Borne, G. (2018). *Surfing and sustainability*. Routledge.

Berkes, F., & Ross, H. (2013). Community resilience: Toward an integrated approach. *Society & Natural Resources*, *26*(1), 5–20.

Copus, C., & Clark, A. (2020). *Local government in England: Centralisation, autonomy and control*. Palgrave Macmillan.

DeVille, N. V., Tomasso, L. P., Stoddard, O. P., Wilt, G. E., Horton, T. H., Wolf, K. L., … & James, P. (2021). Time spent in nature is associated with increased pro-environmental attitudes and behaviors. *International Journal of Environmental Research and Public Health*, *18*(14), 7498.

Ehrlich, P. R. (1968). *The population bomb*. Ballantine Books.

Garcia, E. J., & Vale, B. (2017). *Unravelling sustainability and resilience in the built environment*. Routledge.

Geels, F. (2011). The multi-level perspective on sustainability transitions: Responses to seven criticisms. *Environmental Innovation and Societal Transitions*, *1*(1), 24–40.

Grin, J., Rotmans, J., & Schot, J. (2010). *Transitions to sustainable development: New directions in the study of long term transformative change*. Routledge.

Lowndes, V., & Gardner, A. (2016). Local governance under the conservatives: Super-austerity, devolution and the 'smarter state'. *Local Government Studies*, *42*(3), 357–375.

Macnaghten, P., & Urry, J. (1998). *Contested natures*. Sage.

Magis, K. (2010). Community resilience: An indicator of social sustainability. *Society and Natural Resources*, *23*(5), 401–416.

Martin, L., White, M. P., Hunt, A., Richardson, M., Pahl, S., & Burt, J. (2020). Nature contact, nature connectedness and associations with health, wellbeing and pro-environmental behaviours. *Journal of Environmental Psychology*, *68*, 101389.

Meadows, D. H., Meadows, D. L., Randers, J., & Behrens, W. W. (1972). *The limits to growth*. Universe Books.

Neumayer, E. (2012). Weak versus strong sustainability: Exploring the limits of two opposing paradigms. *Weak versus strong sustainability*. Edward Elgar Publishing.

O'Neill, B. C., Kriegler, E., Ebi, K. L., Kemp-Benedict, E., Riahi, K., Rothman, D. S., … & Solecki, W. (2017). The roads ahead: Narratives for shared socioeconomic pathways describing world futures in the 21st century. *Global Environmental Change*, *42*, 169–180.

Scott, M. J., Gallent, N., & Gkartzios, M. (Eds.). (2019). *The Routledge companion to rural planning* (p. 265). Routledge.

Shucksmith, M. (2018). Re-imagining the rural: From rural idyll to Good Countryside. *Journal of Rural Studies*, *59*, 163–172.

Shucksmith, M., & Brown, D. L. (Eds.). (2016). *Routledge international handbook of rural studies*. Routledge.

Smith, A., Stirling, A., & Berkhout, F. (2005). The governance of sustainable socio-technical transitions. *Research policy*, *34*(10), 1491–1510.Sovacool, B. K. (2016). How long will it take? Conceptualizing the temporal dynamics of energy transitions. *Energy Research & Social Science*, *13*, 202–215.

Tomaney, J., Pike, A., & Rodríguez-Pose, A. (2010). Local and regional development in times of crisis. *Environment and Planning A*, *42*(4), 771–779.

United Nations, Department of Economic and Social Affairs, Population Division. (UN DESA) (2022). *World population prospects 2022: Summary of results*. United Nations. https://population.un.org/wpp/

United Nations Development Programme (UNDP). (2023). *Human development report 2023/2024: Breaking the gridlock – Reimagining cooperation in a polarized world*. United Nations Development Programme. https://hdr.undp.org/

Wheeler, R. (2014). Mining memories in a rural community: Landscape, temporality and place identity. *Journal of Rural Studies*, *36*, 22–32.

Wills, J. (2016). *Locating localism: Statecraft, citizenship and local democracy*. Policy Press.

7 The Salience of Risk

Uncertainty about Global Warming

The previous chapter focused on issues that related to sustainable development. It was a deliberate omission at that point not to associate that discussion with climate change as climate change represents a significant narrative within the data. In Chapter 5, survey data were explored that presented respondents with a structured format for responding to relevant areas relating to climate change and global warming. And in Chapter 6, we once again saw the close association of the narratives of sustainable development and global warming within the interviews. The following discussion will therefore draw out a number of themes as they relate to global warming and are contextualised by the previous chapter on sustainable development. At the outset, it is pertinent to recognise that discourses of sustainable development often included discussions of global warming, but there was no clear causal correlation between the two discourses. For example, a strong understanding of, and discussion in relation to sustainable development did not necessarily indicate a clear understanding of global warming. Also it is worth reiterating, as identified in Chapter 5, that the term global warming is used in this book, from terminology used in the survey through to subsequent discussions and analysis. As already identified, this is to avoid confusion between ozone depletion and the greenhouse effect (Stamm et al. 2000), as well as the term being identified as evoking stronger public concern (Whitmarsh 2009)

Councillors who participated in the interviews were explicitly asked to articulate their understanding of global warming. As with sustainable development there was an element of climate scepticism within the interviews. The following interviewee presents a representative message. *'I do not have sufficient information to make an informed opinion. My own feeling is that there is a lot of scare mongering about it and there's a lot of misinformation coming out on both sides'.* This interviewee identified a number of consistent elements within their narrative. Firstly, there was a lack of information from which to make an informed decision. The role of increased information and education relating to global warming was a consistent theme through the interviews. But more than the lack of education, the interviewee also indicates that scaremongering is an element of the global warming narrative. We can at this point draw on Mike Hulme (2009, 2023) who has discussed global

DOI: 10.4324/9780429297199-7

warming and climate change within the context of 'climates'. Hulme explains that the climate change narrative has been elevated to a global risk that is now unable to be contested, resulting in the subjugation of all other issues. How this is interpreted can be seen through two lenses within a risk society. The first is the role of catastrophe in creating a barrier to effective change and representing processes of counter-reflexivity. This uncertainty can then undermine local action to global warming and also broader issues in relation to sustainable development, which could be said to reinforce what Beck (1992) describes as manufactured uncertainty.

The second perspective is the emancipatory element of catastrophic phenomena such as global warming (Beck 2016). Through this lens the scale and urgency of climate risk can move individuals and communities to look beyond these uncertainties as inaction becomes inconceivable. Even the misinformation the interviewee refers to, paradoxically can be seen to be part of a broader reflexive process. This presents itself through contradictions between scientific consensus on global warming and public doubt, which as it becomes more visible compels society to explore more robust mechanisms for truth and transparency. These issues become even more pertinent today with the onset and rapid evolution of generative artificial intelligence which can both accelerate climate misinformation and present a platform for checking and verifying information (Crawford 2021).

And indeed within the interview the scepticism was tempered by a strong sense that humanity was responsible for global warming. In the following excerpt, the interviewee draws out a number of important issues in relation to humanity's role in global warming.

> Well I think humanity is responsible for global warming, I didn't up to a short time ago, the more recent information has made me feel that we probably are responsible for global warming. My view is still that the amount of money that is being spent because of global warming isn't really justified by the problems that it's going to cause as far as I can see at the moment.

Accepting that humanity was responsible for climate change, the interviewee acknowledges that their opinion of the causes of global warming had changed. This statement was contextualised by a comment that referred to a 'suspicion of scientists'. This is a central narrative and highlights the relationship between scientific and lay knowledges. The interviewees continue.

> I'm afraid that I do have the view that scientists very often put forward views in order to obtain funding for themselves, many scientists have many different ideas but there was a programme recently that showed how you could predict temperatures and how temperatures had been predicted in the past relating them to natural events like the volcanoes, I think there were some intensive sunspots and that sort of thing, and they could predict but suddenly they say it cant explained by anything natural other than what man is doing so that has made me think again.

The way that the issues are communicated forms a central part of this interviewees' understanding of the risk of global warming and the way that it is being addressed. The interviewee explored the impact and extended impact of global warming especially in relation to sea level rise:

> I mean for instance if the sea levels go up, yes people may have to move, other people will have to build their new houses and this will provide employment which we desperately need just now in a time of economic crisis, so I think there's a lot of swings and roundabouts but the measures that are being taken about global warming are enormously expensive and one wonders how they compare with actually the cost of the global warming itself.

Here there is a reference to the connection between the environment risk of sea level rises and the complex economic and social ramifications of these impacts. In this instance there is a reference to migration because of sea level rise as well as the stimulation of certain markets, in this case construction. Whilst this is unsubstantiated and reductionist from a scale of the impact point of view, it does highlight the impression that global warming and investment relating to it can be generative with new form of social and economic advantage. Interestingly, the Stern Review (2006) cautioned against this short-term mitigation suggesting that the long-term costs of inaction far outweigh any immediate economic gains (Stern 2006). Moreover, this observation engages with broader political tensions as to whether global warming should be interpreted as an investment. Extending this logic again we can see familiar discourses of sustainable development that are mediated by immediate economic necessity which is in tension of long-term, intergeneration sustainable development needs. The interviewee also relates the local context to the global situation:

> I mean I appreciate for certain communities global warming is going to be a problem, you know whether the Dutch can make their Dykes a little bit higher or whether these islands in the Pacific are going to be swamped all together, but on a worldwide basis I still think that there's perhaps a little bit too much emphasis being given to it.

The interviewee's narrative acknowledges both the seriousness of the risk of global warming but diminishes their own feeling of being threatened by the impact of global warming through a 'geographical transference'. References to the need for adaptation measures in Denmark as well as reference to small island states in the Pacific illustrate a de-prioritisation of the risks of global warming to this interviewee. Furthermore, interviewees pointed towards the convoluted communication of global warming narratives.

> I don't know, looking at it logically you get so many different pictures where you see and you talk about ice melting and then you have a weather condition which is the reverse of that, you've got terrible cold winters, we've just had

> a really cold winter, I haven't known a winter as cold as it's been since I've lived here, twice before I think it was since we've lived here, so global warming don't sound to be right does it somehow you expect the whole globe to go up half a degree or something, or a degree, and it melt the icecaps but we're not facing it here somehow, I'm not sure about it.

The interviewee articulates the tension between their lived experience and the global scientific discourse of global warming. This discourse does not contradict the scientific evidence of increasing global temperatures, but it does demonstrate that the relevance of these events fails to permeate the life worlds of local residents and their everyday lives. This poses a potential barrier for the development of effective policy and programmes targeting the impact of global warming. But there were many contradictory voices within the interviewees that highlighted the confusion and lack of salience of the global warming narrative. The reference to 'looking at it logically' points to what Wynne (1996) describes as the active role of the public in developing their own understanding of science, which can ultimately lead to ambivalence when narratives appear contradictory.

If we connect these observations to a community level, there was, both within single interviews and across a range of interviewees, a broader tension that mediated through discourses of community. Whilst often at the individual level, the interviewee was sceptical of global warming when the conversation turned to broader community dynamics there was evidence of a shift in this perception towards concern and solution. *'We tend to feel it is a concern to the community but were doing as much as we can, what else can we do?'* The interviewee identifies global warming as a concern for the community but exhibits frustration when it comes to putting any form of tangible action in place to address the related issues. Any solutions that were expressed inevitably focused on the parish context and the mechanisms that were available to address the issues.

> So if we are looking at new properties, we make sure we introduce measures into the planning, when we see planning of a large roof can we do something and say that can we have a solar panel on and things like this.

This presented itself throughout the data where initial scepticism, confusion at one level was replaced with concern for community and an attachment to existing and high-profile solutions. We can see here with reference to the regulatory framework of planning and the technological solutions produced by renewable energy. Reference to solutions also drew on broader initiatives. *We make sure that we look after recycling and those sort of things and please use the buses and not your car and those sort of things, we do say this. But you see we get hit with this transport business.* Transport emerged repeatedly as an issue associated with global warming, on a number of occasions this was also associated with the carbon footprint, though there were broad-ranging narratives. Also, embedded within the discussions on solutions to global warming at the parish council level was the restricted financial resources to be able to address issues effectively.

> Fairest way, um I liked the head tax, I think that was a thing that shouldn't have been stopped, I think that was one of the things and I would have liked to have seen the head tax brought in again and a portion of that because you know it's a human thing and if you tax that human being, if you use a part of that taxing to help a human being I think that's the way to go.

This interviewee explores the notion of the head tax better known as the poll tax or community charge. This was a form of local taxation introduced under the Thatcher government. The tax amount was set by the local authority but was abolished in 1991 when the current council tax system was established.

In Chapter 5, we saw a survey result that addressed a number of global warming-related issues. This ranged from causes of global warming, the relationship between global and local aspects of global warming, to issues that may be said to impact on global warming at the local level. These issues included the housing stock, affordable housing, flood plain, settlement types, transport, and the inefficient appraisal of greenhouse gas emissions. And to lesser or greater degrees, these issues all presented themselves within the interviews. The following discussions will build on these initial observations from the baseline of perceptions as to whether climate change is actually happening at all, if it is happening is it completely a natural phenomenon, a natural phenomenon but being made worse by humans, or completely caused by human activities on the planet. Figure 7.1 reminds us of the survey results in this regard.

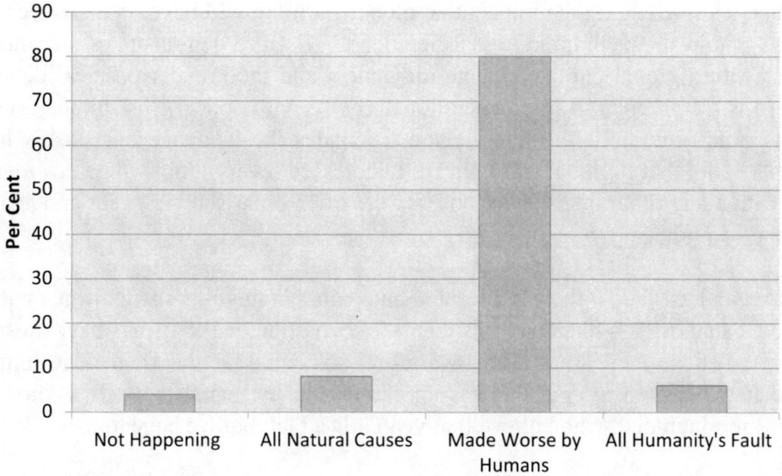

Figure 7.1 Causes of Global Warming

It is clear from the survey that the majority of respondents indicated that they felt that global warming was made worse by human actions on the environment. Interview data supported this finding but also offered a more sophisticated understanding of the processes that are involved in the construction of narratives at the individual and community levels. Interview data also offered a deeper insight into the relationship between the risks from global warming and the relationship that individuals and communities have with those risks.

> There's no doubt that man has contributed to it since probably the start of the industrial revolution around about 1700 when industry really got going. Its not the only cause and its international.

This interviewee expresses a clear perception that humanity has contributed towards global warming. There is also a recognition that this was set in motion since the industrial revolution and the beginning of the modern era. However, these comments are also tempered with an understanding that global warming is 'not the only cause' and that it is international in context. The interviewee also articulated a narrative that addresses sceptical positions.

> Well I know it's happening and the sceptics are the ones who are totally opposed to it who I think are like ostriches who bury their heads in the sand, Al Gore's got it right, I think what we've got is unless we tackle it, our children, our grandchildren, will end up with a planet that's going to be destroyed.

There are multiple processes evident within his excerpt, processes that are also evident across the corpus of interview material. The interviewee diminishes climate sceptics' positions by claiming that they are burying their heads in the sand, by not acknowledging the impact that global warming will have on their lives and those within their communities. Reference to Al Gore demonstrates an engagement with a global climate change movement and the broader political debates. There is a reference to intergenerational equity with concern for future generations. And ultimately, the interviewee articulates the language of catastrophism with a belief that if global warming isn't addressed 'everything will be destroyed'. As with many other interviewees there were often polarised views for and against global warming.

> I mean lets' face it there's a for and an against in this life, there always will be, I suppose it's like Darwin's theory of evolution, we'll always have those who believe there's not global warming, this could just be a transient thing that will pass way, but there is an element in the parish council, I know, myself included, who believe that yeah it is a fact that we have to face.

This statement exposes the recognition that different perspectives are a fact of life, normalising a polemic perspective, and indeed this is so engrained in the human psyche that it is part of an evolutionary process. Also, this discourse frames both global warming and Darwinian theory as subjects of enduring disagreement. Building on the previous observations, this situates global warming within the contested domains of science, where empirical evidence is filtered through cultural, religious, and ideological lenses. Furthermore, identifying global warming as *a fact we have to face* points to an important reflexive dynamic, suggesting that global risks are not dependent on belief, their material consequences necessitate social and political responses. Other members of the parish council also feel that global warming is an issue to be addressed within the remit of local government. The interviewee continued to explain that the environment committee within the council was tasked with exploring these issues and they were linked to tangible action within the parish.

> I think like the cycle racks, encouraging people to save energy, we've got our sort of save energy bulbs, yeah there is, and of course we're collecting batteries as well for recycling, we're doing all the green bits, so yes I'm very pleased with the way we're going on that.

Issues of transport, through enabling more access to cycling through cycle routes, and energy use, conservation, and production are brought to the fore in the conversation. As outlined above, there were inevitably differing views, some polarised, some demonstrating a more discursive and less absolute positions towards the reality or otherwise of the global warming phenomena. Some interviewees' activity expressed an abstinence in making a decision one way or the other.

> I am sitting on the fence because I think some things people say are global warming are things that have happened in the past and its just a natural cycle of things. I'm not really sure about that at all.

The natural cycle narrative was dominant for those councillors that were sceptical or uncertain. This narrative extended to the community scale. Interviewees acknowledged that other priorities inhibited their own and members of their parishes' ability to be concerned about global warming. '*There's many more things that are more concerned, where's the next wage packet or the next meal or yeah I think there's more important things*'. Again, the financial imperative of individuals, families and communities takes precedence over the broader concerns of global risks. This tension is well articulated in the following statement from an interviewee. '*I mean the term being banded around is I don't know whether to heat or eat. And there are a few like that*'. This same interviewee indicated that they did not feel that there was anything that their parish council could do to address the risks and consequences of global warming.

The discussion so far highlights a number of issues. It explores how local understandings of global warming are negotiated at the intersection between uncertainty, scepticism, and the demands of everyday life. References to natural cycles demonstrate the way that global risks are reinterpreted through familiar cultural lenses. At the same time others have adopted positions of abstention illustrating a reluctance to commit in the face of contradictory claims. We have also seen how global warming discourses which intimately entangled with the realities of community cohesion which ultimately shapes notions of agency and responsibility. These insights continue to develop our understanding of the relationship between sustainable development and reflexive modernity through interpretations of global and local risk. Moving the analysis forward, the survey and the interviews drew out understanding and representation of carbon footprint as a way of connection the abstract nature of global warming to tangible individual and community responsibility and action. This enables a strong connection to a reflexive modernity. For Beck, risks are often abstract, invisible, and distant from the experience of individuals, forcing discussion on responsibility though narratives of the carbon footprint allows reflexive processes to emerge from the interview process.

When the issue of carbon footprint was raised within interviews, there was an overwhelming indication by councillors that they were not only aware of the concept but that it was an important concept in response to global warming. Interviewees articulated how they understood carbon footprint both personally and more generally.

> I'm the furthest away so I pick up a car load rather than four or five of us taking a car but that's about the only thing that we can do to help in that way. We do, we are aware shall we say.

Carbon footprint for this interviewee is articulated through transport use and a willingness amongst the parish councillors to carpool to the meetings.

> I think its very alarming and I think it is a great concern to all of us in the world and I think that we need to planting lots and lots and lots of trees. And I think we should all be aware of our own carbon footprint and I think it is an issue that we all should take on board because I don't know in a hundred years time what our climate is going to be like here. Whether we are going to have green trees around us, water and I think it is not a flippant thing. I think it is very important.

Here the interviewee expressed significant concern in relation to global warming and this translates into a feeling that there should be a broader awareness of not only the interviewees' own carbon footprint but that everyone should be aware of their carbon footprint. The interviewee also future casts to a period where nature has been significantly impacted by the effects of global warming. This concern for nature and the sense of potential loss of the natural environment resonated strongly with the interviewees, with multiple narratives that explored global warming within the context of a dystopian future.

Environmental concerns were broad ranging from the loss of habitats, coastal erosion, flooding, and increased sea level. These concerns reflected those identified in the survey. There was also evidence of the relationship between the global and local and how this prioritised within the parish council community. The way that nature and the environment are articulated extends beyond the project to a dystopian future where the environment becomes the focus of a negative side effect of global warming. As we have already discussed, whilst this is no doubt a central theme, there is also a version of the future that utilises nature to provide solutions. This has already begun to emerge in the previous discussion. The following excerpt highlights this position:

> Yes, it worries, it may not be in my lifetime but its going to be in our next generations lifetime and I hope that our sustainable energy that we use is not going to be nuclear. I hope its going to be sort of from the sea or wind power or and I think that we have very, engineers and people working on this are now and I hope that in the next future there's going to be an energy that is going to be safe. A safe environment.

Initially, the interviewee acknowledges the extended timelines by saying that solutions may not be available in their own lifetime. Again, there is a consistent emphasis on future generations. The conversation then turns to sources of sustainable energy, and whilst there was a strong presence of this narrative throughout the interviewees, the interviewee makes clear distinctions between desirable energy sources and desirable futures. The use of nuclear power was seen as an undesirable future and not an option for Cornwall or a globally sustainable future. Renewable energy from the sea and wind is highlighted as the most appropriate sources of energy. Again, the interviews overwhelmingly emphasised the unique landscapes throughout Cornwall. With turbines being situated in rural open areas to maximise energy production means that in certain areas turbines dominate the landscape. The scale of the turbines ranges from small community projects to larger commercial wind farms. A number of wind generators are community-owned and involve significant community investment. Table 7.1 identifies the key areas of wind generation in Cornwall.

Table 7.1 Sites of Wind Generation in Cornwall

Area/Location	*Details*
Penwith Peninsula	Includes the St Just and St Ives areas. Notable for various smaller wind farms and individual turbines.
Bodmin Moor	Home to several wind turbines, particularly in areas like Davidstow Moor.
North Cornwall	Includes sites such as the wind farm near Delabole, which is one of the oldest onshore wind farms in the UK.
Camborne and Redruth Area	Features turbines on the outskirts, contributing to local renewable energy efforts.
Goonhilly Downs	Known for the Goonhilly Wind Farm, which benefits from the open landscape and strong winds.
St Austell Bay Area	Includes various smaller turbines and community-owned wind projects.
Truro Area	Features a few local turbines contributing to the city's renewable energy resources.

Table 7.2 Community-Owned or Locally Invested Wind Projects in Cornwall

Project Name	*Location*	*Details*
St Breock Wind Farm	Near Wadebridge	Originally developed as a commercial project but has seen local investment and involvement.
Lizard Peninsula Community Wind Farm	Lizard Peninsula	A community-owned project involving local residents in ownership and decision-making.
Solarcentury's Community Wind Farm	Near St Austell	Includes local investment and a community share offer to allow residents to participate in the project.
Cornwall Community Wind Farm	Various locations in Cornwall	A community-led initiative where local residents can invest and benefit from the profits.
Goonhilly Downs Wind Farm	Goonhilly Downs	Whilst not entirely community-owned, it has involved significant local investment and engagement.

There are also a number of smaller wind generation sites that are community-owned in Cornwall. Table 7.2 outlines these sites.

Carbon footprints were repeatedly connected to renewable energy and wind turbines. But this narrative took on different forms that provided insights into the interviewees' construction of the topic. The following interviewee demonstrates this process:

> I'm not totally opposed to trying to get rid of carbon as such, but then the carbon that is coming out was all there before, now unfortunately we're upsetting the balance slightly by doing certain things and releasing it early, now if we can cut that down fine, and we can, electricity can all be generated totally carbon-free but then we've got the do-gooders who would object to certain things.

The underlying discourse is framed through an assumption of equilibrium. Inherent within nature is a balance and it is the disturbance of this balance which leads to unintended and negative impacts. In this instance carbon is naturally occurring and naturally regulated within the natural global cycles. And it is humanity premature releasing of this carbon which has led to the negative side effects of accelerated global warming. This narrative is married to the narrative of solutions through the production of carbon neutral energy. There is also a glimpse in this excerpt of a view of those that are opposed to carbon neutral pathway as understood by this interviewee. The term 'do gooder' is used in a derogatory sense to diminish the view of those whose views are in contrast to those of the interviewees. Variations of this narrative can be seen throughout the interviewee data. This narrative is not necessarily tethered to a pro or con of renewable energy or broader sustainable development-related issues, instead it is used to rationale the interviewees own perspective as they develop their views during the interview process. The interviewee continues to validate their own views in contrast to the 'do gooders' as they relate

Table 7.3 Wave Energy Projects

Project Name	*Location*	*Details*
Wave Hub	Hayle	A significant test facility for wave energy technologies. It provides an area for developers to test and demonstrate their wave energy devices.
FaBTest	Falmouth	The Floating and Benthic Test facility for wave and tidal energy devices. It offers a test environment for marine renewable energy technologies.
Plymouth Marine Laboratory	Plymouth	Whilst not a wave energy project per se, the lab conducts significant research into marine energy technologies, including wave power.
Wave Dragon	Near Falmouth	A prototype of a wave energy converter designed to capture and convert wave energy into electricity. Although not deployed in Cornwall, it has been tested in nearby waters.

to wind turbines. *'I mean why they object to wind turbines, I think they're elegant, there are certain places for not putting them, historically I would say but not otherwise*'. The aesthetics of the wind turbines are described as elegant and there is no reason, if appropriately placed why wind generation should not be welcomed.

With the longest coastline of any county in England, wave energy was also highlighted as a viable source of energy. Whilst not as widely deployed Cornwall has a number of wave energy projects. Table 7.3 highlights these.

The interviewees' narrative concludes with safety and the need to feel safe and minimise risk. The interviewee continues to articulate global warming beyond their own perspective and addresses whether they felt it was of broader concern to their community.

> I don't know. I think its all down very much to the individual and how they look at it. I mean if it was a topic that was blown up and people then maybe it would be. I think at the moment its all a personal point of view really.

However, there is a reinforcement in the interviewees' narrative of the role that the parish council plays. Whilst the initial comments highlight that the community might not be concerned about global warming and personal viewpoints differ greatly, the parish is seen here as an enabler for responding to the risks of global warming.

> I think the parish council should be looking at the carbon footprint definitely. And I think it is an area that perhaps we should take on board and also our recycling and that. But the thing about recycling at the moment everybody is recycling but China's not taking it.

For the interviewee there is a fluidity in the conversation between perceptions, issues, and solutions. This emerged repeatedly as the interviewees sought to articulate their views and understanding of global warming and sustainable development more broadly. Here the narrative moves from carbon footprint to recycling as a natural connection for the interviewee within the context of global warming. Furthermore, there is a recognition and broader understanding of the global dynamics that exist within the global warming narrative. In this instance the narrative recognises the barriers that can be imposed through the actions of country's that are embedded in the broader global networks of recycling. The interviewee is referring to the collapse of China's biggest importer of waste in 2009. There is a direct connection between the global dynamics of the international waste and recycling industry and that identified within Cornwall.

> I mean when this all was very much recycle, recycle, recycle there was no problem. It was all going over to China and now we haven't got anywhere. We've got loads and loads of recycling and its got nowhere to go. We've had a big issue over a site in Cornwall to be and its been stopped. Actually it's a very very big issue.

The interviews then create a sophisticated and often contradictory narrative of the way that global warming is understood and responded to. The interviews demonstrate that the lived experience of parish councillors, their parishes, and their communities connect in different ways with the global narratives and information flows. As the interviews progressed and more and more narratives were explored the richness of the perspectives on global warming continued to emerge. As we have already discussed, and as is evident as we build depth to the landscape of narratives explored whether global warming is anthropogenic or not continues to emerge. The notion of cyclical events presents as a strong theme as is demonstrated in the following except. I *think global warming is cyclical – if you look back in history and when they're digging up the ice and measuring these things, you'll find these cycles in the ice, they found them already.* The interviewee draws on a scientific discourse to validate their perspective. Using examples of ice core records, this interviewee's view is rationalised through expertise but seems to lack the specific context. This expert knowledge framing is then connected very directly to the local context and lay knowledge understanding of global warming.

> And as for warming round here, I think we have gone up about 1 degree but we're still getting awful cold winters, although not as bad as the freak winters, we had one 12, 15 years ago something like that, but you're going to get freak weather anyway from time to time.

Personal experience of freak weather is used to support the view that global warming is cyclical. Combined with ice sheet records this creates a powerful rationalisation for the interviewee's perspective.

This narrative in the interview is then extended to the parish council members as a whole and related to any action taken. There was a particular focus on whether or not parish councils were active in reducing any perceived risk from global warming. This varied considerably amongst the parish's. *Not directly, I mean 1 or 2 of us will travel in the same car, which obviously reduces it. We try to do that but no nothing directly that we do.* These narratives were often combined with the way that global warming was communicated amongst parish members. There is often a tethering of narrative to other issues that don't directly address global warming as a central concern. *No, it is mentioned in planning when somebody asks for a turbine; it's only mentioned in passing basically.* And where this is again extended to the parish's community areas, the perception is that little action is being taken. *Not that I'm aware of. The odd person's bought a turbine mind, but then that's purely to keep the electricity bill down.* This statement is present in different ways throughout the corpus of interviews. Again reversion to a connection with renewable energy is a common mechanism for understanding global warming. Solutions were often confused and emerged as conversations progressed. Inevitably, legislation was seen as a prominent vehicle for addressing global warming and often this was associated with a particular policy area. The following demonstrates this in relation to transportation:

> Well yes we ought to introduce some form of legislation to get more back on the railways, we should extend the railways and their collection depots, there ought to be far more collection depots. The country was the best served in the world for rail transport, now it's all very well saying

Sitting on the fence and uncertainty were key themes in the corpus of interviews. It is unsurprisingly an organising framework within the narratives presented. For and against, uncertain as well as abstaining from formulating an opinion at all. And whilst this is a fundamental recognition of the climate change discourse, both at the time the data was collected and views that are evident in multiple forums today, it is the underlying motivations that provide the richest insights into engagement with climate change. On a number of occasions, interviewees expressed a level of excitement that they were being given the opportunity to really discuss these issues and there was an unfolding of their understanding of the issues. The following interviewee demonstrates this well:

> Right I've got mixed views, I've watched the programmes on it, I've listened to the speakers on it and commonsense tells me that there are things that I can do which if you believe what the experts are saying, could help to combat GW and the bad affects of the human population on the planet, so that's fine.

The interviewee indicates that they have received information on global warming from multiple sources. These sources function as the conduit through which the messaging of global warming is received and internalised. Television and speakers feed into a context that is understood as common sense. This common sense understanding is also met with the expert opinion which is either to be or not to be believed. Again it is worth highlighting the consistent emphasis within the corpus of data that there is a strong relationship between the narratives of expert information and lay knowledge. This relationship is not definitive and whilst in some cases it is mutually reinforcing when it comes to the global warming narrative, in the vast majority of discourses it is identified as a tension what manifests through conflicting world views. Returning to the previous interviewee the ambivalence towards global warming begins to be expressed.

> But at the same time I think to myself why isn't our government doing more, why don't, they're supposed to be sitting there, why don't they do things, for example we spend a lot of time doing the recycling here, we have recycling bins put down round there by the cottages, we encourage all our guests and ourselves obviously, that we recycle plastic, paper, cardboard, glass and tin, that's what we do and it's my job to sort it all out and because some of it we don't have collections for it, I have to put it in the back of the car and take it to the recycling centre.

The interviewee demonstrates key behavioural traits in relation to recycling but simultaneously acknowledges the need to use their own transportation which could present a behavioural contradiction. This is in relation to holiday cottages that they rent to tourists. Interviewees often included elements of their personal and extended activities beyond the parish council role which enabled them to construct a robust understanding of the issue being discussed. Here there is a convergence of the personal, business, and parish council roles that highlight the complex and often contested ways that perceptions of global warming are articulated. These dynamics underpin the overt rhetoric and conscious articulation of the global warming narrative by the interviewer.

This observation in of itself presents a key issue for providing multi-scale solutions to complex, uncertain, and constantly reconstructed issues. These narratives also reinforce broader contentions within a risk society that there is a need to accommodate multiple knowledges that converge between scientific, indigenous, and the extra-experiential. And whilst we unpick the route source, or sources of individual narratives and perceptions, at the surface levels interviewees must grapple with the contradictions to make sense of their position within them. The following excerpt is built on the previous observations as the interviewee articulates frustration at programme and policy solutions that don't appear to address the reality of community needs and plastic usage.

> But what is so annoying about it is that we have a kerbside collection that'll only collect the plastic bottles, they won't collect the other bits of plastic, and I was saying to people that I know on the recycling thing for Cornwall, and they're lovely people, they're doing a real good job, they come along to the farmers' market and put a stand up, is why is it that they're using plastic that the people don't use.

The previous excerpt highlights the analysis by drawing together a number of competing and reinforcing narratives. Initially, there is the infrastructural element of only being able to collect plastic bottles and not other forms of potentially recyclable plastics. This extends to a dichotomy between the value systems of the organisations and the people that are implementing the schemes. In this instance they are 'really nice people' and the inability to provide a more rounded solution to recycling multiple plastics. Again, we can draw on the underpinning processes of knowledge formation to shed light on this perceived dichotomy within the infrastructure and human perspectives of recycling. There is perhaps a lack of understanding by the interviewee of the scientific processes for recycling plastics. However, the same interviewee continues to offer a more sophisticated understanding of why only bottles can be recycled in that context.

> Now surely the government are in a position where they could say you will only use this category of plastic, so it's going to be recycling, and they could give a timescale, they could give the supermarkets time to you know and manufacturers some time to implement that particular type of plastic, but they don't so, so much of the stuff is kind of rejected.

As is demonstrated in the previous extract, the interviewee highlights the relationship between government, business, in the form of supermarkets, and the technological process of manufacturing. As this interviewee progressed, the level of sophistication attached to the narrative also increased and allowed the interviewee to make connections across multiple domains. The interviewee goes on to discuss behavioural change and especially within the context of financial incentives. '*We don't want that, so I think well if you're serious about it why don't you do that, and if you're serious about providing other things then encourage people to do it, you know they made money available for the banks haven't they now but for goodness sake'*. The interviewee continues to make further connections identifying the ways that a parish council can act as a catalyst for change in response to global warming. Another interviewee also discusses the role of financial incentives.

> If you were to address this global warming situation by encouraging people to do what we do with the heat pump or with solar heating or wind turbines or things like that in areas where it would work, say well look give a grant, give a grant to people to do it, that saves them some money, it helps combat global warming and it actually creates some jobs for someone to actually manufacture the stuff, a big win couldn't we, but they don't seem to have any sort of strategic thing.

Here the interviewee highlights the climate mitigation measures as catalysts, socially and economically, global warming is framed not only as an environmental necessity but also as an opportunity for economic growth, innovation, and job creation. The reference to grants and financial incentives points to the importance of governance structures in enabling this transition. The reference to a lack of strategy things potentially points to the interviewee's frustration with perceived policy gaps as well as an absence of a coordinated responses. In tandem with this the interviewee links environmental action with local benefits. Understood in this way his extract resonates with broader sustainable development discourse directing the reader to the integration of social, environmental, and economic benefits. And throughout the interview, there seemed to be an undercurrent of frustration in relation to the effectiveness of governance processes at the local level for responding to global warming.

> I find with our guests, they come from different areas around the country so they recycle things in our bins in a different way than we do because in their areas its different, their council will accept this in their waste whereas ours won't, curious, why isn't the government taking the lead and saying look do you know what I mean, would it be that hard to appoint someone, recycling supreme, now tackle it, what glass, what thing, let's set up the infrastructure, don't send it off to China in a barge, for goodness sake deal with it.

This councillor is drawing attention to the inconsistent and fragmented nature of environmental governance at the local level. But in this instance it is not the frustration at the public's willingness to act, but instead the lack of coherent national leadership and effective infrastructure. The reference to 'recycling supreme' points to the piecemeal and inconsistent approaches that local authorities take, that essentially undermine collective responsibility as well as eroding trust. And once again buried in this narrative we identify a core dilemma within a reflexive modernity. That is, global risks require coordination and transnational responses, yet governance responses remain fragmented and nationally and locally dependent. But, at the same time, the reference to 'don't send it off to China' indicates a broader unease with globalised systems of responsibility where solutions are abstract and imposed instead of locally relevant and cocreated.

Building on the observation and frustrations of the previous councillor in relation to national leadership and fragmented governance, another interview discusses organisations at the global scale.

> Organisations in the sort of global sense, organisations are slow to catch up. I think we've got to the point now where we need to think about how we are going to deal with the future not about trying to stop it happening because its happening and the rate at which globally, the timescale on which global events operate right, well put it like this.

Referring to global organisations, the interviewee claims that global organisations are slow to catch up and this is in the context of adaptation to global warming and not focusing on mitigation because 'its happening'. The reference to extended timescales for this interviewee presents tensions between global and local responses. This also emerged within the interviewees in different contexts adding another layer to the global local dynamic. The discourses within the interview represent these multiple global elements: material, cultural, political, psychological, economic, and risk. Global warming draws these elements together emphasising it as a wicked problem. The multi-scalar frustrations seem to demonstrate how community level discourses of sustainable development are not only about local action and personal action, but also about a perceived vacuum at different levels of governance. A core principle of a risk society is institutional inadequacy in the face of globalised risk. The following interviewee's excerpt seems to embody this institutional inadequacy quite succinctly:

> Its like a super tanker right heading towards a port about five miles away right and you think oh, I'd better turn the engines off but its going to coast isn't it? Its going to keep going and whatever you do, even if you've got everybody outside with paddles trying to stop it, you ain't going to stop it. Its going to happen. In six months time whatever by the time it gets there so in a way global warming is a bit like that right.

The analogy of the super tanker highlights again the inertia, the relationship between action and impact and inevitably raised questions over the viability of responses and actions. This interviewee continues to explore this line of thought by referring to the 'genie being out of the bottle' and the processes of global warming have already been set in motion. This sits within a form of environmental determinism which, it could be argued actually reduces collective and individual agency. We witness this form of determinism in broader societal narratives on global warming that draw on terms such as tipping point, scientific consensus often related to calls to action, internationally and locally. Phrases such as tipping point have become representative of global warming debates within broader scientific media and policy arenas (Russill & Nyssa 2009). Similarly, the use of scientific consensus catalysis calls to action at different levels (Oreskes 2004; Van der Linden et al. 2015). The rise of the climate emergency movement is representative of these narratives with the rhetoric being utilised across sectors to catalyse political and individual commitment (Davidson et al. 2020; Howarth et al. 2021). This leaves individuals with a behavioural dilemma as to how to act for themselves and their communities, and this presented itself throughout the interviews. The following excerpt illustrates this very well:

> Because on the global scale of things and the timescale of things its happening and what we need to think about more now is how we're going to adapt agriculture in this country, how are we going to adapt housing and our lifestyles and so forth. How the health system is going to cope with a possible influx of malaria if the temperature rises. I mean we are already seeing malaria move up from Africa you know, how do we know what to do in the face of all that.

This interviewee connects global and local perspectives and suggests that timescales are short to be able to address the complex and uncertain risks. The excerpt further reflects behavioural dilemmas where uncertainty and perceived lack of individual and collective efficacy can limit engagement (Lorenzoni et al. 2007). There is a recognition that the potential risks form global warming require significant, if not systemic change to infrastructure and lifestyle. This ranges from housing to health systems and presents a strong reflection of the impact of global narrative.

But there is also clear uncertainty as to how to act in the face of these risks and the impacts that they will have. Another interviewee highlights the need for global policy to focus on adaptation.

> So I think what we in a global sense should be concentrating our efforts on now, is how to live with the mess we have ended up creating. Don't bother about reducing gases. You can do that if you want but you ain't going to change the fact that its happening.

This level of fatalistic, 'it's too late' rhetoric did not dominate the local narrative but it was clearly present. And when related to the parish community there was a sense of good intention but actions having minimal impact.

> I think, yes, first of all for the community yes there is an awareness. People do have a sense of global warming and so forth and they will, they will talk about, they will do their bit. People will do their bit but in a way they feel their well intentions are too wide of the mark.' This interview also highlighted the negative impact that a sense of 'doing the right thing' could have. 'A lot of people compost at home so a lot of people do their thing but in some ways they are a bit misguided. Its like the people who load up the car with all their bottles to take to the bottle bank and drive there to drop the bloody things off when they weren't going out for anything else.

Whilst these narratives point to action, there is also an undertone of helplessness leading to disengagement, even with the awareness of these risks (Borne 2009; Norgaard 2011). At the community level, the well-intentioned environmentally motivated actions such as composting and recycling are acknowledged but framed as misguided and insufficient. This seems to resonate strongly with Stove's critique of the attitudes, behavioural, choice model (ABC), where climate policy does not sufficiently include meaningful systemic infrastructure shifts for genuine change. Taking these observations into account, the previous discussion has attempted to highlight a range of narratives from calls for practical incentives such as grants for renewable energy to critiques of fragmented recycling systems and the absence of national and global strategies. The constant relationship between the global and local scales is ever-present as we see issues of institutional inertia, inconsistent infrastructure, and the potential ability of cumbersome global organisations to undermine trust and limit effective responses to risk. And whilst some have framed global risk as an opportunity for innovation and even job creation, others adopt a more cautious and fatalistic approach, seeing adaptation to risks such as global warming as the only option.

Conclusion

This chapter has explored the salience of risk and builds on the previous chapter which has, in one way or another developed a continued narrative of parish councillors relationship with risk. What continues to emerge is a deeply layered and contradictory narrative in which scientific uncertainty, lay knowledges, and local realities intersect in complex ways. Whilst many of the interviewees accepted that global warming was an important issue, a substantial number of interviewees expressed scepticism especially relating to scientific messaging, expert motivations, and the incomplete and uneven communication of risk. Clearly, these observations reflect the broader reflexive modern dynamics of the tension between lived experience and local expert systems of knowledge. For councillors, the discrepancy between global narratives of catastrophe and their own experiences created a sense of confusion and disengagement.

But despite doubts, there is a strong sense of community-level concern about the environment where many interviewees identified tangible, localised actions, some of which included, recycling, promoting public transport, installing solar panels, and supporting community-owned renewable energy. Based on this observation it is reasonable to suggest that there is an active though fragmented engagement with sustainable development. We also see that this engagement often bypasses formal discourses and occupies a space of place-based action.

At a practical level, the interviewees identified structural and procedural barriers to local climate action. These included the limited capacity within the councils themselves, the ever-present lack of government support as well as unclear communication relating to responsibility which ultimately leads to frustration. But these points were tempered by interviewees' strong support for the ability of local government, under the right circumstances to be effective mechanisms for community improvement. Interviewees felt that parish councils had key roles to play in raising awareness, setting behavioural examples, and essentially catalysing change. We can relate this directly to indicators of reflexive modernity where local actors engage in a process of experimentation to create adaptive solutions, even in the face of global uncertainty. And indeed global uncertainty and a perceived disconnect between the global and the local level priorities were a recurring theme. Whilst some councillors did internalise global risks, others relegated them to distant locales with little relevance for the daily life experience of Cornish people and communities. But entwined with all of these narratives was a desire to see more connection between national policy, global strategy, and local implementation.

In summary, data presented in this chapter point to the urgency for engagement with the culture, emotional, and pragmatic dimensions of global risk. The assumption of a straightforward and linear dissemination of global policy to local implementation and practice is seen as ineffective. Instead, what is required is a complex and reflexive process of engagement that is underpinned by trust, local knowledge, personal values, and recognition of institutional capacity. This is the space that a reflexive modernity occupies.

Bibliography

Beck, U. (1992). *Risk society, towards a new modernity*. Sage.

Beck, U. (2016). *The metamorphosis of the world: How climate change is transforming our concept of the world*. John Wiley & Sons.

Borne, G. (2009). Achieving sustainable lifestyles or encouraging a counter-reflexivity: Exploring motivations for sustainability in a mediated risk society. *Local Environment, 14*(1), 93–107.

Crawford, K. (2021). *Atlas of AI: Power, politics, and the planetary costs of artificial intelligence*. Yale University Press.

Davidson, K., Briggs, J., Nolan, E., Bush, J., & Håkansson, I. (2020). The making of a climate emergency response: Examining the attributes of climate emergency plans. *Urban Climate, 33*. Article 100666.

Howarth, C., Lane, M., & Fankhauser, S. (2021). What next for local climate emergency declarations? The gap between rhetoric and action. *Climatic Change, 167*, 27.

Hulme, M. (2023). *Climate change isn't everything: Liberating climate politics from alarmism*. John Wiley & Sons.

Hulme, M. (2009). *Why we disagree about climate change: Understanding controversy, inaction and opportunity*. Cambridge University Press.

Lorenzoni, I., Nicholson-Cole, S., & Whitmarsh, L. (2007). Barriers perceived to engaging with climate change among the UK public and their policy implications. *Global Environmental Change, 17*(3–4), 445–459.

Norgaard, K. M. (2011). *Living in denial: Climate change, emotions, and everyday life*. MIT Press.

Oreskes, N. (2004). The scientific consensus on climate change. *Science, 306*(5702), 1686–1686.

Russill, C., & Nyssa, Z. (2009). The tipping point trend in climate change communication. *Global Environmental Change, 19*(3), 336–344.

Stamm, K., Clark, F., & Eblacas, P. (2000). Mass communication and public understanding of environmental problems: The case of global warming. *Public Understanding of Science, 9*(3), 219–237.

Stern, N. (2006). *The Stern review on the economics of climate change*. Cambridge University Press.

Van der Linden, S. L., Leiserowitz, A. A., Feinberg, G. D., & Maibach, E. W. (2015). The scientific consensus on climate change as a gateway belief: Experimental evidence. *PloS One*, 10(2).

Whitmarsh, L. (2009). What's in a name? Commonalities and differences in public understanding of "climate change" and "global warming". *Public Understanding of Science, 18*(4), 401–420.

Wynne, B. (1996). Misunderstood misunderstanding: Social identities and public uptake of science. In A. Irwin & B. Wynne (Eds.), *Misunderstanding science? The public reconstruction of science and technology* (pp. 19–46). Cambridge University Press.

8 Sustainable Communities

Community Unpacked

It's useful in the first instance to discuss the way we understand community within the context of this book. We have already explored notions of community in relation to parish councils identifying a sub-political space of assemblages and nexus points of flow and network, and we take this understanding forward in this chapter. I have previously identified the notion of community as a highly ambiguous concept with reorganisation of social space in previous decades adding to the diversity and fluidity of interpretation. Within a risk society, the traditional notions of community as static, bounded, and place-based are disrupted by global and local risks. We have already drawn this out in previous chapters as we explored the discourses of sustainable development, global warming, and risk. Throughout discussions so far, we have witnessed the dependence of parish councils, their communities, and the individual lies within them on wider, political, economic, and knowledge systems.

However, we must also acknowledge that community remains a powerful and highly relevant category in many ways. For identify, problem, and solution framing, for example. For the practical job of service provision. For creating meaning and mobilising resources. Communities are the assemblages where the abstract nature of global risk is translated into the lived experience of the members of that community. Centrally to the issue presented in this chapter, the aforementioned tension creates responses that can be defined by ambivalence, scepticism, and competing priorities, all of which we have already witnessed as we have explored the narratives presented by town and parish councillors. Extending these observations within the context of a risk society, it is reasonable to suggest that community operates as both sites of vulnerability, exposed to risks that they cannot control, but also sites of reflexivity, where awareness of global interdependence can catalyse new responses and practices to the imperatives of a sustainable development.

Drawing on the foundations of a sustainable development, the notion of a sustainable community is diverse and contested, reflecting the social, economic, and environmental priorities of multiple actors as well as shifting political and social priorities. We saw in Chapter 3 that within the context of the Sustainable Communities Act 2007, sustainable communities are positioned within the themes of local

DOI: 10.4324/9780429297199-8

economic, environmental protection, social inclusion, and democratic involvement. UK government policy has variously pointed to sustainable communities as places where people want to live and work and where the diverse needs of current and future generations can be met, again we see how this resonates with definitions of sustainable development (ODPM 2005; WCED 1987). The Cornwall Plan 2020–2050 frames sustainable communities as those that are greener, cleaner, safer, and also referring to affordable housing infrastructure. With that in mind we begin by contextualising a broad understanding of community through the survey results. In line with earlier discussions on globalisation, assemblages and the relationship between the global and local spheres of analysis, the survey sought to understand community beyond a unit of geographical space and explore ideas of multiple community attachment from the respondents (Figure 8.1).

Councillors were asked a number of questions about their community that was designed to elicit a sense of belonging not just to a single community but to multiple communities that exist in a single geographical area. Logically the first question that was broached with respondents was whether they felt part of a community. The vast majority of respondents indicated that they did feel part of a community at 95 per cent. The following questions were designed to understand whether town and parish councillors felt that they were members of more than one community. It was of little surprise that the majority of respondents at 65 per cent indicated that they were part of more than one community. At the same time overwhelmingly, 89 per cent said that their communities were in their local area. Here, these tensions of simultaneous geographical location and de-territorialisation sit comfortably with our concept of community at the sub-political level as we have already discussed. They do not sit as comfortably with conventional policy narratives that often exist in the two-dimensional world of administrative boundaries.

With the initial tension in community construction exposed, we now move to explore the interviews which provided a rich and varied understanding of community, sustainable community, and the risks that were associated with their

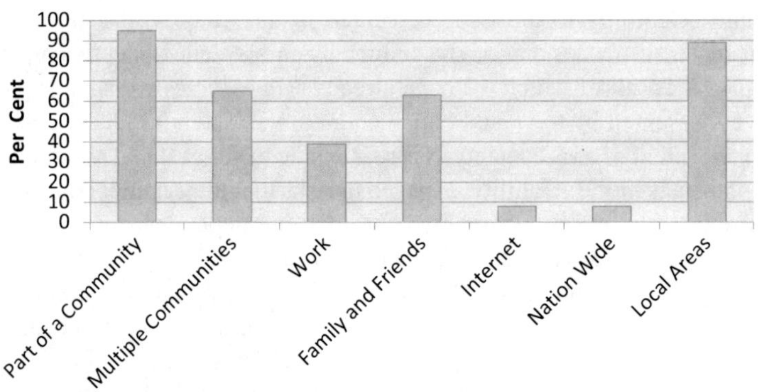

Figure 8.1 Community Dynamics

communities. As with all elements of the interview data, the number of issues that presented themes in the context of community was vast. In order to address both the breadth and depth of issues raised the following approach is taken. Initially, there is a deep dive into a small number of select interviews that set the scene for the complexity of the various narratives, themes, and perceptions that emerged. There is then an interwoven narrative that moves to identify a cross-section of key themes that emerged from the data drawing on multiple interviews and extracts from the corpus of interviewee data. There are also summary tables that collate the themes, provide a brief explanation of that theme and connect to the chapter's overall analysis.

We begin unpacking the narrative of a single interview. The following interview demonstrates the broader spectrum of issues that were represented in many interviews in relation to this topic.

> Well I think again it's about bringing back like I said for having that wish list really, I mean I don't know, in theory it would be great wouldn't it to have local businesses back there and sort of a thriving community that way so that communities are less reliant on supermarkets and things like that, and that knocks down all the environmental issues as well with travelling and packaging and stuff like that, but again it's about cost and if you've got families on a budget they can't afford necessarily to use the local shops so that's sort of Catch 22.

Immediately, there is a reference to local businesses and the relationship this has to a thriving community. In contrast large supermarkets are identified as the antithesis to this goal. Furthermore, the interviewee identifies multiple environmental issues that would be addressed through this local lens. This included transportation and packaging, but this narrative is tempered with the economic reality that faces many families identifying the catch-22 scenario that faces many families. The interviewee continues to contextualise their perspective.

> You'd want to think that all those services were there so they wouldn't need, it would create a sustainable so you'd actually have more local employment back into the town, you wouldn't have to travel too far to actually get employment. So, I would look at it from that point of view trying to make it sustainable but also sustainable in the sense of the community involvement through the volunteering so actually getting younger people to get involved with local projects, local events and keeping the community spirit of the town as well.

Here, the focus shifts to the services that are provided within the local area as well as relating this to local employment. However, the interviewee then expands this narrative to move beyond the economistic dimension of community sustainability to highlight the importance of 'community spirit' through local community engagement and volunteering. This is a consistent theme throughout the interviews, where there are attempts to grapple with the tangible realities of community cohesion and the necessities of economic reality that drives the conversation of broader policy and political engagement. A community spirit also highlights how sustainable development is not only about mitigating risks but also preserving social identity and cohesion. With reference to sustainable development this resonates with broader discourses that stress the importance of social capital and civic engagement.

The framework of the narrative progresses, and we witness the active construction of the interviewees' deep engagement with their local community. This unfolds through the lens of community responsibility.

> I think also again looking at community responsibilities, so looking at areas of need within the area, what needs to be maintained, what things could be developed, what things could, you know we're looking at allotments well that could involve families, children getting involved in that as well, taking on some of the, growing the vegetables that sort of thing, so it's about looking at the bigger picture rather than keeping it quite insular, I think that does tend to happen as much as we say we've got a community spirit, which we have, people are still very focused on, I've got to pay the rent, I've got to do this and not actually looking at things.

The contractions and tensions persist in the previous excerpt as the interview identifies the important role of community responsibility, the involvement of family groups and networks around food and allotments, but cautions against an insular perspective that misses the bigger picture and the broader national and global dynamics that impact local areas. As well as the previously identified tension, that transcends multiple areas in different contexts, the desire for a community spirit and the necessity of paying the rent. As the conversation progressed, the relationship with sustainable development and the broader parish council governance began to emerge. The interviewee indicated that sustainability wasn't necessarily a topic that was raised during the parish meetings which often had a very tight agenda dealing with the everyday business of parish council activities. The active nature of the interview process forced the issues which the interviewee at that time had not considered and became a catalyst for sustainability discussion beyond the environment of the interview.

> I think it would be interesting to hear, certainly from some of the questions you've asked me, it would be interesting to hear other people's points of view on what they think is sustainable, cos a lot of them with respect don't really care.

The conversation continued to touch on multiple consistent themes that exist throughout the corpus of interviews.

> I think one of the problems just for example, of travel, cos that's probably the major problem for me, I either work here or I work in the village and no

> matter how much I say well I ought to catch the bus, just as an example of that, is difficult to do.

Recycling was a consistent and easily identifiable environmental behaviour that many interviewees identified within their lexicon of environmental behaviours. And this recognition of the personal responsibility towards recycling often extended to a recognition that more could be done to encourage other members of the community to engage in recycling activities.

> I could do with encouraging more people to do it as well, and just like the general affects of environment, I think we need to get more people involved in that and I think we almost need to, because I wanted to ban the carrier bag in the town but obviously we've got loads of supermarkets in the town, to do it was a difficulty.

Here we witness the emergence of the consistent relationship between aspirations towards environmental behaviour and the structural and systemic barriers to achieving this behaviour. This observation is a consistent theme not only throughout the data presented here, but also in the broader literature and research. And whilst not a revolutionary observation, the scale of the data collected strongly supports and resonates with the existing studies pointing to a significant and enduring barrier to transitioning to more sustainable ways of operating (Blake 1999; Hoffmann et al. 2024; Kollmuss & Agyeman 2002).

Another strong theme that emerged from the data wasn't only about behaviours towards the environment and narratives of what a sustainable community may mean, but also the impact of the natural environment on behaviours themselves, on community perception of the world and the sense of connection to the world. *Yeah, it's a whole thing, it's a package isn't it with you know the environment, I mean a lot of this is areas of outstanding natural beauty and these fields down below had that status.* Resonating with discourses identified in Chapter 6, here there is reference to the natural environment and in particular the classifications of the land, which is tempered with concerns over developmental processes within these areas.

There was often recognition that local communities were just getting by with an inevitable exodus of younger generations moving for the aforementioned economic imperative to find work and broader life opportunities. This inevitably reflects the consistent gap between the ideal of a sustainable community and the reality of achieving this goal. It is more broadly recognised that outmigration is a significant barrier to achieving a sustainable community. There was also a recognition that there could be limited social interactions for younger generations which potentially could lead to a sense of social and geographical isolation. As we have seen previously the interviews touched on broader risks associated with global warming and global climate change. There was a mixture of engagement with the narrative from media representations through to direct perceived engagement from extreme weather events. But there was an overriding narrative that communities at large were not engaging with the narrative in a meaningful way. That is not to say that there was a strong sentiment of the importance of the issues relating to climate change but the principle focus of parish council work which addresses the practicalities of local issues, planning, and infrastructure needs inevitably override these broader topics. There is no significant focus on global environmental issues like global warming or carbon footprints, which suggests a need for broader environmental education and integration of sustainability into local governance.

Broader themes that were explored in the interview included funding for environmental projects, where consideration was given to funding sources and where the responsibility for this funding should be placed. Whilst initially the interviewee considered funding for environmental projects should come from central government through taxation, they acknowledged the difficulty in raising money from the community itself, reflecting a reliance on government intervention for large-scale initiatives. Inevitably the effectiveness of parish councils in general was explored. Consistent with many interviewees, there was an acknowledgment that the parish councils' role is pivotal and it is an effective agent within the local community. There was a recognition that the parish council fulfils its obligations and responsibilities within its scope of activities especially in relation to planning and infrastructure issues. Interestingly, the interviewee saw the parish council's role as reactive instead of playing a more proactive role in terms of sustainability and environmental issues. The interview reflects the broader corpus of data as communities in Cornwall grapple with multiple pressures relating to rurality, infrastructure, tourism, and economic instability.

An Expanded Narrative

The previous section has drawn on a small number of representative cases in order to deeply explore the evolving narratives of the interviewees. It is essential to recognise the interview process as an active event, something that helps catalyse in the minds of the interviewees the issues being discussed. As such, there is a lack of linearity in the conversations that were held. And whilst interview prompts were used to guide the conversations and attempt to address particular issues, there was an inevitable meandering of personal reflections and an interplay between introspective examinations and external engagement with the natural and social worlds. This only added to the richness of the data collected, which yielded multiple themes and points for broader discussion. In the following section we explore themes more broadly across the interviews.

The importance of the involvement of community members in decision making, as well as including key actors, was repeatedly illustrated. 'We really make an effort to involve everyone in the planning meetings… it's important that all voices are heard'. This sentiment was encountered on multiple occasions with a lack of participation, consultation, and involvement in decision-making processes identified as a barrier to effect governance. '*The intention is everyone should be involved, but its often the case that certain interests are presented and others are just silent, that's just life I suppose*'. And in combination with this there was a consistent reference to gauging the opinion of the local community and associated organisations.

'What would be better would be to seek ideas from the rest of the community, and I think part of our role and our obligation is to make sure that we create a space where that is possible'. Fostering sustainable communities was often articulated in terms of livelihoods. 'It goes without saying that if there aren't any jobs in the area if people can't support their families then, probably very reluctantly they will have to leave the county, we all know this happens all the time'.

This narrative was frequently combined with broader discussions around place identity and especially in the context of resistance to change. *'Moving out of Cornwall if often a real struggle for a lot of people it's about heritage, connection, the village is the very soul of the person, it's difficult for people from cities and larger towns to understand that'*. This observation was further embellished with three parallel narratives. One related to a cautious, if not resistant approach to development as well as sustainable development measures, especially those that were externally imposed. The view of 'imposed' solutions challenged individual and community values. Some interviews expressed concerns that there was a fear of losing the distinctiveness and social cohesion of small communities. This obviously presents a tension even a paradox of perceived loss of community cohesion because of policies designed to protect community cohesion.

One interviewee said that '*some just want to keep things as they were in the 1950s'*, suggesting that there is scepticism about anything modern. And many others referred to the unique character of Cornwall which has *'...always been a place apart there is that Cornish feeling and people just want to protect that'*. And more broadly the sentiment that Cornwall isn't '*...just a place to visit it's a place to live'*. But there were also voices within he interviewees that presented the second parallel narrative which recognised that there wasn't a polemic, a tension, or indeed the paradox as previously described. A number of interviewees communicated a future where heritage and tradition coexisted with innovation and carefully thought through and inclusive development. 'Theres a difference between tradition and being stuck. We've got to make sure we don't confuse the two'. Other interviewees echoed this sentiment. '*You can honour the past, your family and background, without hindering progress and the need for change'*. Others give tangible examples of where small communities have had to respond to modern needs and changing circumstances. *'When I came to living in this village, we just had this little village institute, now we've got community halls, a youth club, a surf club, all built up over time. That's progress, but rooted in the village spirit'*.

And these narratives persist in different contexts, they varied from parish to parish, from small to larger council and represented the richness of peoples lived experiences. Whilst there is significant scope to expand on this issue perhaps the following interviewee sums up the overarching sentiment of many councillors. *'I'm a traditionalist in many ways, but you can't fight everything. You've got to pick your battles and move forward together'*.

These connections to place were also frequently connected to nature. The third parallel narrative which repeatedly intertwined with many representations of community was that of green and blue spaces, connection to the Cornish landscape and the need to balance that connection with conservation and access. *'It's a lovely place to be... we haven't got sort of a major man made attraction, but there's a strong sense that the countryside is, well it draws you in'.* Interviewees often expressed a profound emotional connection to their local environment. This emerged from interviewees that had a strong multigenerational family connection to Cornwall as well as people that had recently moved to the county. *'I'm not Cornish by birth, but the land draws you in. You become part of it. It's not just a place to live; it becomes part of who you are'*. Here we see the clear sense of identity that the landscape has for this interviewee. Not Cornish by birth but profoundly impacted by the natural environment. There is an intertwined element of this narrative that acknowledges a breakdown of the barrier between the self and nature. And this sentiment presented itself on numerous occasions contributing to the overall understanding community sustainability and cohesion. And inevitably this connection extended to behaviours towards the environment. '*There's just something about this land it holds you. I think that's why people fight so hard to protect it.'* Whilst responses towards the environment differ, this idea of protection was central to a personal articulation of nature and an extension of the dissolution of boundaries between the self and the natural environment.

Protecting the environment is seen as synonymous with protecting the self. The following excerpt further reinforces this point: *'People move here for the lifestyle, the peace, the beauty, the coast. And that's something we have to protect'*. These articulations of nature are connected to multiple areas of community cohesion and interpretations of what a sustainable community meant to participants.

Running alongside the previous observations, the importance of ownership of land and nature presented a consistent theme in the data. Notions of access, exclusion, and the relationship between public and private ownership presented in various forms but existed as a strong sentiment throughout the research. The following example highlights the relationship between public and private ownership and the impact this has on social and economic factors:

> We've had the woodland handed over to us, a public amenity really. Somewhere where you can walk, have picnic and stuff. When it was private, it was just left. The youngsters would go in there and perhaps be drinking and possibly drugs and lighted fires. We thought if the parish council took it over we'd have perhaps more control.

The woodland area is identified as a community asset, essential for recreation, wellbeing, and part of the nature connection, is also identified as being a site for negative social activities, drinking, drugs, and further antisocial behaviour. The overall sentiment presented here is that the parish council through community ownership can both connect and preserve the natural asset of the community, and positively contribute to reducing social bads through that ownership.

Other interviewees pointed to the important role that nature plays in areas where economic activity has declined. Again this relates to the notion of nature as a free asset, as something that can and should be accessed by everyone. *'The pub is closed, but we've got the countryside, and you can walk right from your door. I think that's something people value even more now, having that connection'*. With the removal of a central community asset in rural areas, the local pub, the connection and access to nature is seen here as heightened in importance and essential for the continued sustainability of the community. Many more examples of the way the natural environment impacted interviewees' interpretation of community and sustainable community are evident. Within the data these emerged through broader themes such as emotional and intergenerational bonds to the landscape. '*I was born here, I live in spitting distance of that field and I certainly wouldn't leave it*'. This interview expressed a deep emotional connection to a field that had been part of their family for generations. '*I hope my children's children are able to enjoy it, I hope the house is still here*'. These short excerpts represent the connection for deep emotional ties to the land and a sense of custodianship. The identity and distinctives of place repeatedly emerged through reference to Cornishness. We have previously discussed the distinctive nature of Cornwall beyond the geographical boundaries. Perhaps the following excerpt encapsulates this sentiment that was visible throughout the interviews. '*There is this Cornish feeling, Cornwall has always been a place apart*'. And with this in mind, the following discussion highlights an issue that forms a central theme throughout the councillors' narratives of Cornwall.

The Ocean

Whilst over a third of parish councils in Cornwall have boundaries that directly border the sea, there are significantly more councils that have close proximity to the ocean, and even more that are connected directly through estuarine and river processes. Table 8.1 highlights the number of town and parish councils that have boundaries with coastal or estuarine and river systems, as well as the number of parish councils that may be considered inland.

To a certain extent, understanding connection to the ocean builds on the observation identified in Chapter 4 as to whether respondents to the survey felt their parish council was rural or urban. And whilst I expected the ocean to be a strong theme, through my own connection with the county and the ocean, especially through the activity of ocean surfing, the interviewees presented dimensions of this attachment that I had not considered.

> The sea gives us everything and takes it away just as quick. You learn to live with that rhythm. But that's what binds us. We're not just building houses,

Table 8.1 Parish Type

Parish Type	*Number*	*Total (%)*	*Notes*
Directly coastal (bordering the sea)	85	39.9	Atlantic and Channel coasts
Estuarine/major river (Fal, Tamar, Camel, Fowey, Helford, etc.)	42	19.7	Strong maritime/ocean identity but not open coast
Inland (no direct sea/estuary/river frontage)	86	40.4	Still reference the ocean in community identity
Total	213	100	

> we're trying to keep the coast walkable, keep the beaches clean, protect the view that people come for and that we grew up with.

Extending the connection to the local environment, there was a clear and consistent theme that often dominated the conversations within interviews. This relates to the ocean, the Cornish coastline, of which there is 433 miles, and the deep maritime heritage of Cornish social structures and the economic impact this has had on the county. This emerged as a key theme for community identity, governance, sustainability, and personal identity.

> Cornwall, the people who live in Cornwall and I stress the difference, I'm not talking about specifically the Cornish, people who live in Cornwall think of themselves as different, the fact that we're surrounded on three sides by sea and on the other side by the River Tamar is very much part of the ethos of the whole county… it's the identity of the people who live here and I feel it too.

This interviewee articulates how Cornwall's identity is shaped by the ocean. The ocean is seen not merely as a resource to be exploited but as a physical boundary that creates distinctiveness, belonging, and cohesion. The references to the county's physical separation with the county being surrounded on three sides by the sea, and then severed from mainland England by the River Tamar. This separation then plays into the nationalist rhetoric within Cornwall and functions as both a cultural and emotional container. And this deep cultural connection to the ocean underpinning the broader economic importance that stabilised communities, created livelihoods, and built entire communities. *'Yes of course for fishing and tourism, but for more tenuous reasons, the sea is an icon, it's the identity of the people that live here'.* And here in lies a glimpse of the need for governance, policy, legislation, and any form of regulation to be sensitive to, and considerate of the ocean identity of Cornish communities. Policy initiatives that disregard, or try in some form to decouple community from their maritime identity, imposing any environmental or economic intervention that marginalises the role of the sea, runs the risk of not only being ineffectual, but also undermining a community's sense of belonging.

And these persistent themes of belonging and ocean connectedness were balanced with a move away from a romanticism of the natural environment towards a pragmatic recognition of the risks that are inherent within these fragile natural areas. Even where interviewees identified as not being coastal, the impact of the ocean and fluvial processes was still prevalent within the discourse.

> We're not right on the coast and even on the river, with it being hilly, most houses are up above the river, but there are big issues in the river, there's a natural environment to protect, boats being abandoned and rotting in the river.

It's possible to draw multiple themes from this statement. There is an overt environmental stewardship that is a pragmatic response to visible degradation. This interviewee continues to frame their narratives of sustainability through custodianship and a need to manage and preserve the local environment in the face of regulatory gaps and behavioural neglect. And references to action on the ground both through local policy and individual behaviour were repeatedly connected in the littoral zone as this councillor indicates.

> We are very keen and the whole ethos of this parish plan was that we, well our parish environment is not only about how we protect and preserve the natural features but also about how we manage the manmade challenge… Yes, that's why we put a recycling unit on the beach. Local issues like that we are trying to encourage people to keep the place clean and tidy.

Here the role of the parish plan is identified as a mechanism to preserve the natural environment and respond to the impact in the sensitive coastal zones. The establishment of beach cleaning units provides an infrastructure to engage local populations and tourists in a meaningful and visible manner. Another factor that emerged from direct action, such as community beach cleans, is related to how these events can become the focal point of broader sustainable governance activities. These activities engage broader stakeholders and partners such as local businesses and environmental organisations. There was also recognition of some of the barriers that hosting these events presents.

> The parish council used to put up and organise a voluntary beach clean, but now it's come down to more health and safety. Should you wear rubber gloves? should you be supplied with long handled clipper things to pick the debris up? We don't clean the seaweed away. That's natural. It's only plastics and bottle tops and fishing nets.

And whilst health and safety is identified as potentially presenting a barrier to the expansion of these activities, these events still provide an effective mechanism for engaging the community whilst simultaneously providing a sense of community cohesion focused on environmental protection that promotes civic action. There is also an interesting reference to the relationship between seaweed and plastics as natural and non-natural litter, which could present interesting avenues for future investigation, especially in the context of new materialist discourses. Plastics are seen as out of place, disruptive, and rubbish, whereas seaweed is natural and where it belongs. Yet both materials are shaping ecological practice and community identity. Here we witness socio-ecological assemblages as well as material politics by deciding what does and does not constitute natural materials. (Barad 2007; Bennett 2010; Latour 2005).

Contextualised within a risk society we can articulate plastics as environmental hazards of modern processes. The resultant action and reflection by local communities, through beach cleaning in this instance, represents a reflexive modern response where local actors must critically negotiate the consequences of industrial modernisation. Moreover, these actions represent a sustainable development that highlights both the challenges of managing environmental risks and the opportunities for catalysing responsibility and civic responsibility.

Other issues that occurred throughout the interviews related to the interconnection between water quality, marine ecosystems, and broader coastal health. This councillor explains that

> The river runs onto the beach and we have the water quality check and the water quality notices go up every week. In high season we have three volunteer beach cleans out of season which involves being sponsored by local businesses.

Again there is a connection to broader stakeholders and the role of local business in creating solutions-based coalitions. As mentioned previously, the maritime history of Cornwall is central to identity, community cohesion, and past and future narratives of the county. Fishing inevitably emerged as a central activity within the county and it is important to elaborate on this here. Throughout the interviews, there were discussions about the decline of Cornwall's fishing economy. But, as with the decline of any major economic activity in an area, this decline extended beyond the economic impacts to highlight a shift in meaning, identity, and the rhythm of everyday life.

Fishing the Tides of Change: Coastal Identity and Community

For generations fishing defined the meaning and identity of community in Cornwall with the challenges, risks, highs, and lows that this precarious lifestyle engendered. Interviewees reflected not only on the industry decline, but also on the decline in identity and purpose. Whilst there is a strong temptation to delve very deeply into this specific topic, the following will draw essential connections across the interviews and connect to the key issue identified in this chapter. This is achieved through four interlinked themes. Firstly, the decline of the fishing industry and working coasts, secondly, the fragmentation of communities, thirdly, the effect of second home ownership, and fourthly, the enduring spirit and responsibility of coastal stewardship. Whilst many of the councillors talked about fishing anecdotally, many also had direct family and friendship connections to the industry. The following councillor also had a career as a fisherman:

> I was a fisherman for 28 years before I changed over, the fishing industry's dead here in the village now. When I was fishing there was 20 of us, now we're down to 2. We had the slipway here, now it's been busted up by bad weather. It's like the sea's still there, but it's not part of life in the same way anymore.

This statement epitomises the industry decline and the impact that it has on communities throughout the county. And this statement represents not only a loss of infrastructure and economic value, but the loss of direct and lived coastal knowledge. The 'busted slipway' becomes a very emotive representation of the disconnect that the decline of a central regional industry has on community life. And whilst this sentiment was consistent throughout the interviews there was also a recognition that fishing still has a role to play in community life. *'There's still a good lot of fishing community, but we've lost a lot of it… and that will go too unless we can keep the youngsters there'*.

This response highlights the continued importance of fishing but recognises the significant barriers that exist for the continued existence of the industry. The fishing industry's survival is recognised in time scales that are generational being contingent on being able to retain younger generations in the area to ensure a viable workforce, as well as the continuation of embedded knowledge and heritage connections. And in other areas of Cornwall, interviewees discussed how traditional fishing communities had reinvented themselves to respond to the fluctuations in economic activities.

> The docks have reinvented themselves. They've got some contracts with the Royal Fleet Axillary, the Navy, cruise liners also play a big part now. One of our regeneration schemes is dredging the harbour because some ships can't get alongside anymore.

The example of this parish's connection to the ocean presented a strong reimagining of the future of the docks for future economic prosperity. This was possible because the existing infrastructure had the unique physical characteristics of the coastline in that area. Even in the two brief statements already presented, we see the common themes of fishing decline, but the significantly uneven response that this has within a relatively confined geographic space. Nevertheless, both interviewees point to the importance of maintaining and revitalising Cornwall's maritime heritage.

As the interviews unfolded and more and more perspectives came together, the role of the ocean and its connection to community in Cornwall developed in different directions. A reality of not only Cornwall but also many coastal communities throughout the country, and indeed the world was the recognition of coastal isolation and the fragility of rural cohesions.

Coastal Isolation

Interwoven with the importance of fishing to Cornish identity councillors also reflected more broadly on how the sea shapes, or indeed fails to shape social cohesion within their communities. The smaller parishes isolated and disconnected locations in certain coastal areas seem to magnify and accelerate the impact of the decline of the fishing industry, and was then combined with other economic realities that compounded the disintegration of local communities.

> Its heartbreaking but I think its safe to say that these are communities literally coming apart at the seams. They are very small and very far apart, there's no school anymore, the chapel is closed and a great number of people that I talk to say there are isolated and lonely. Its not a very connected place even though we are surrounded by all this history and countryside and the sea.

This sentiment juxtaposes the heritage, history, and raw beauty of the countryside and the ocean, arguably the essence of Cornishness, yet that is a vacuum within which local communities exist that is defined by absence and not presence of social cohesion, a consistent representation of cultural decline. With that said there were, within close geographical proximity, parallel perspectives that represent a different narrative. '*We've got a rugby club, cricket club, football, British Legion, shops, a library... the council's very active, lots of things going on*'. This contrast is a continuing reminder that we are not exploring pockets of uniformity; we are exploring the full tapestry of human life and the divergent and inconsistent narratives that this presents us with. In this instance whilst both the parishes from which these narratives are drawn are close to the sea, their trajectories clearly diverged. The former parish identified a hollowing out of the services and distinct community features, whilst the latter community retains its community institutions and protects its civil momentum.

There are then spacial and aspacial factors that impact the reality of community cohesion and point to dimensions of prosperity and decline. The aforementioned notions of coastal isolation or prosperity are not necessarily dictated by the proximity to the ocean but more ingrained realities of rural life. And even in parishes that encompass larger towns there are mixed narratives of affluence and decline. And this councillor of a larger Cornish town indicates.

> There aren't that many opportunities for social mixing, the retail economy's skewed towards tourists, it's not like a proper market town anymore where everyone came in to do their shopping, chat about their lives and sit on the front and just enjoy the view.

And here we witness a consistently emerging narrative referring to community and localism in states of flux. A shifting presentation of community away from a functional localism to something more performative and responsive that necessarily responds to the increasing realities of tourism. If it is fair to argue therefore that in this transitional space the sea is no longer the centre piece of community life, drawing people together in a meaningful deep connection way? What we witness perhaps instead is a connection to the ocean that emerges through consumption, both materially and through entertainment and leisure activities.

We can very easily contextualise the previous observations with a risk society and the unintended consequences of modernisation. Combinations of industrial decline, demographic changes, the rise of tourism-driven economies, all create new vulnerabilities that communities must address through reflexivity. As previously observed in this context the ocean is no longer a unifying entity but is reformed through patterns of leisure and consumption as well as tensions between heritage and globalised economic pressures. Furthermore, framed within sustainable development, these dynamics underscore the need to balance cultural identity, economic adaptation, and social cohesion in ways that can both preserve Cornish distinctiveness and secure viable futures for its coastal and inland communities.

Second Homes Tourism and the Displacement of Belonging

Whilst we have already discussed the ever-present spectre of the impact of second home ownership within Cornwall and rural and coastal communities more broadly, it is virtually impossible to divorce this reality from the narratives of the ocean within the interviews. The impact of second homes and seasonal tourism is one of the most pervasive themes in relation to coastal communities, so we revisit this theme here within the context of ocean and coast. Whilst these issues were often framed within the context of economic opportunities, or lack thereof, this was often inexorably connected to displacement of local residents and the disruption of community. As this councillor indicates.

> You've got a split community, people holding things to the past and then a new community, What you get is a lot of 'we don't want this', we don't want that' — meanwhile, people can't get a house or park their car.

There are two pertinent issues incumbent in the previous excerpt. Not only does the rise in second homes make the housing market beyond the reach of local residents, but there is also the emergence of a political inertia where planning decisions can be stifled by preservationist nostalgia. '*We've got a lot of incomers, quite affluent people, they like the peace and quiet, but they don't come to things in the village hall'*. This results in, at the very least, a two-tiered coastal experience. On the one hand, the local residents struggle to create a place-based identity, and on the other second home owners, recent residents, in this case 'newcomers', exist in a space of detached consumption. But of course, this two-tiered approach is intertwined in many instances, and the narratives and goals of broader policy are enshrined by attempts to draw on one to benefit the other. This hybrid narrative of mutual coexistence was evident throughout the interviews. '*We've got a large student population now, lots of regeneration money, the waterfront is being integrated into town life, we're trying to manage the pressures of second homes with active engagement and planning'*.

Here the council has identified opportunities and positioned itself as a coastal curator of available resources. There is an overt acknowledgment within this interview that the risks from tourism are very real, but by being proactively engaged in managing these risks through regeneration strategies and community planning, a balance can be struck that is mutually beneficial. And whilst the curator approach is designed to enable effective transfer and application of resources through governance processes, there remains a sense of community ownership in coastal areas that adds to the complexity of place-based solutions. So just as the decline of the fishing industry exposed how fragile coastal communities and livelihoods can be, the pressures of second homes and tourism expose another layer of disruption to community identity, showing how Cornwall's connection to the ocean is continually reshaped by wider economic and social forces.

Community Stewardship: Reclaiming the Coast from Within

Whilst ocean narratives were often about loss this is countered through representations that highlight the rebuilding of coastal life. Through ideas of legacy, local investment, and an ethical approach to development and progress. The following interviewee presents a powerful example of legacy as well as an example of intergenerational stewardship. '*I've already made up my mind to donate a couple of acres for community land. Provide some space for development but only build houses for local people, for local families, we have to keep the youngsters here'*.

Whilst at surface level this statement reflects one individual, with the means and desire to rebuild the community, it also represents a narrative that is innately political. It embodies a narrative and a perspective that is strongly resistant to absentee ownership and a trend of gentrification. This interviewees' approach is that the coast is not lost to the consumerist tendencies that tourism facilities, but instead coastal communities can be empowered to create a prosperous and sustainable future. This sentiment is reinforced in various ways, reflecting a need to support stronger social architecture that facilitates community cohesion. *'We have a town of about 80 organisations, but they don't work together enough. What we need is someone to say, we're all working for the same goal. That's what it needs'*. The following councillor discusses the steps they take as a parish council to create the community voice. *'We run the Art Gallery, we're building a community plan, we've got a resident newsletter, we're trying to move from isolation to inclusion'*. These narratives present personal, political, and institutional perspectives that in some form are talking about reclaiming coastal life within a community context.

Within this chapter we are only able to scratch the surface of the connection between individual, community, and ocean. We have glimpsed the shifting meaning and the material realities of Cornwall's coastal identity through the voices of its town and parish councillors. From the decline of fishing to the rise of second homes, from fragmented parishes to regenerative planning, the ocean is revealed not as a static backdrop but as a living entity, symbolically and politically. And importantly, the interviews point to a coastal sustainability that is more than managing coastal dynamic, erosion, access, regulation, and development. As important as these issues are, they point to a nurturing of belonging, creating an environment of participation, whilst resisting the subjugation of the local voice by more powerful narratives. Throughout the interview, we witness the simultaneous contradictions and inconsistencies of perspectives and action. Whilst some communities mourn the loss of heritage and long-held values, others mobilise engaging the resources at their disposal. And of course even within each community there are fragments of the aforementioned positions. So whilst some slipways inevitably fall into disrepair, others look for alternative connections, alternative access, and alternative opportunities, the commonality of all voices presented here is the intimate relationship to place that is both contested and constantly changing. And whilst the voices that provide these insights are drawn from the closest level to local government, and are overwhelmingly engaged with the role of the parish council, the overarching lesson is that sustainable development, in a remote and fractured geography bounded by the ocean begins not with policy priorities, but with the lived experiences of the people in the communities that the parish councils serve.

Conclusion

This chapter has emphasised the relationship between sustainable development and community. It is recognised that in the context of Cornwall's town and parish councils, this is an intricate and constantly evolving relationship. In line with the conception of parish councils within a sub-political context, and recalling local councils as assemblies and nexus points of global and local flow, community is not a fixed geographical entity, it is a dynamic, multilayered construct shaped by lived experience, socio-economic conditions, environmental connections, and local governance dynamics.

The interviews were able to reveal the lived contradictions of community construction and the tensions that exist. Tensions between tradition and change, between inclusion and isolation, the tension between the romanticised rural identity and the lived reality of economic survival. Whilst councillors recognised community through the lens of physical infrastructure and service provision, there were parallel and more intangible value-driven interpretations relating to community spirit, custodianship, and the emotional connection to land and sea. Indeed expressions of stewardship, both environmentally and culturally, underscore the potential to create locally driven sustainable development.

This chapter has also hinted at the tensions between the aspirations of creating a sustainable development-led community and the systemic barriers that prevent this from being realised. This included practical barriers such as affordability, service access, housing pressures from second homes and a loss of younger generations because of restricted opportunity. Other issues that were identified included planning constraints and the prominence of tourism in the county.

Environmental awareness was strong, with councillors drawing inspiration from Cornwall's natural landscape and unbroken coastline, that inspired action including recycling, beach cleaning, and nature preservation. And at this point we can argue that the notion of a reflexive modernity moves beyond the abstract, beyond the academic, and is directly connected with the lived experience of the interviewees. Parish councillors have to navigate the spectre of global risks but prioritise local needs and they must do this simultaneously.

As with previous chapters we can, at this point, engage with the reflexive modern indicators and overlay the identified narratives. There was clearly a questioning of expert systems, and desire for local priorities and relevant solutions. There was also evidence of the negation of individual identity in a complex and ever-changing world. We have seen that interviewees described the contraction of traditional livelihoods, fishing, for example, alongside attempts to reinvent and reinvigorate coastal life through regeneration and extended partnerships. The ocean was identified both as a resource and a cultural symbol, an ever-present force that anchors Cornish identity and has the ability to catalyse community cohesion. As we have seen, whilst traditional fishing industries have declined the ocean remains central to Cornish life even in the face of coastal pressures. Challenges of over tourism and housing becoming increasingly unaffordable to transient populations creating disjointed and fractured communities in some areas, whilst the same time others manage to maintain strong networks and social infrastructure.

What then can we say about the meaning of community identified in this chapter within a risk society? I will argue that this chapter demonstrates that creating sustainable communities, in a risk society, requires more than a technical infrastructure investment, though this is important, and it requires more than the adherence to environmental protection and regulation. Sustainable communities in a risk society required the ability to listen, listen to the struggles, listen to the joys, and listen to the lived experience of people in these communities. Town and parish councils have the capacity to listen to their communities and translate that into meaningful action and meaningful policy. Parish councils, the most local level of government, if properly engaged, have the capacity to significantly support the future of sustainable development implementation.

Bibliography

Barad, K. (2007). *Meeting the universe halfway: Quantum physics and the entanglement of matter and meaning*. Duke University Press.

Bennett, J. (2010). *Vibrant matter: A political ecology of things*. Duke University Press.

Blake, J. (1999). Overcoming the 'value-action gap' in environmental policy: Tensions between national policy and local experience. *Local Environment*, *4*(3), 257–278.

Cornwall & Isles of Scilly Leadership Board. (2020). *The Cornwall plan 2020–2050*. Cornwall Council.

Hoffmann, R., Kanitsar, G., & Seifert, M. (2024). Behavioral barriers impede pro-environmental decision-making: Experimental evidence from incentivized laboratory and vignette studies. *Ecological Economics*, *225*, 108347.

Kollmuss, A., & Agyeman, J. (2002). Mind the gap: Why do people act environmentally and what are the barriers to pro-environmental behaviour? *Environmental Education Research*, *8*(3), 239–260.

Latour, B. (2005). *Reassembling the social: An introduction to actor-network-theory*. Oxford University Press.

Office of the Deputy Prime Minister (ODPM). (2005). *Sustainable communities: People, places and prosperity*. HM Government.

World Commission on Environment and Development (WCED). (1987). *Our common future*. Oxford University Press.

9 Identifying the Local to Achieve the Global

Sustainable Development within the UN

During an intensive period of ethnographic research, I explored the various ways that sustainable development was utilised in the governance structure of the UN (Borne 2006). At that time I was working for the UN Environment Programme participating in meetings and negotiations of the 57th (2002) UN General Assembly (UNGA) with a clear remit to identify specific areas related to sustainable development. This period of time marked the 12-month anniversary of the attack on the World Trade Centre, and the lead-up to the Iraq war. There was a progressive build-up of tension in the corridors, conference rooms, and main chambers at UN headquarters in New York as diplomatic solutions to the tensions with Iraq began to run out. As part of my role, I was privy to some of the conversations and diplomatic events that took place. The experience transformed my understanding of global politics and the way that the UN grappled with divergent ideologies, political priorities, and economic disparities. It is pertinent to briefly outline the escalation of the situation relating to Iraq during that time.

It was early September when the UN and Iraqi officials resumed negotiations over weapons inspections after a long break in discussions. Ultimately, these talks failed and paved the way for the eventual justification of the war. On 12 September, George W. Bush addressed the UNGA where he called on the UN to challenge Saddam Hussein on suspected violations of multiple Security Council Resolutions. There was a strong narrative from Bush that if the UN did not act, the organisation would become irrelevant on the global stage undermining its entire raison d'etre. Within days Iraq had agreed in principle to the unconditional return of weapons inspectors. On 8 November, the UN Security Council adopted Resolution 1441 giving Iraq a final opportunity to comply with disarmament obligations and weapons expectations preceded und the leadership of Hans Blix.

On 7 December, Iraq officials submitted a 12,000-page weapons declaration categorically claiming that it did not have weapons of mass destruction, this was immediately dismissed by both the US and the UK as being incomplete and obscuring the true situation. As December moved on, the United States continued to accuse Iraq of non-compliance and other members of the security council increased diplomatic pressures arguing that Iraq was in breach of Resolution 1441. Early in

DOI: 10.4324/9780429297199-9

2003, the US and the UK attempted to secure a second UN resolution, explicitly authorising war; this failed, but on 20 March 2003, Iraq war officially began. In September 2004 the war was declared illegal under international law.

So with the context of my initial engagement with the UN outlined we move now to connect these worlds even sot the theoretical underpinnings of this book, a risk society. Applied to Beck's central tenets, the build-up to the Iraq war arguably validates Beck's claim that we are now in a world risk society. This is because the justifications for the war were not necessarily grounded in verifiable evidence-based reality but through a sense of anticipated risk with the possibility of weapons of mass destruction. Beck suggests that the logic of risk displaces traditional evidence-based decision-making, empowering actors to mobilise fear and uncertainty as instruments of political authority. The politics of risk in this instance foreclosed what could have been a cosmopolitan moment of collective security, instead legitimising unilateral action and undermining the authority of the UN.

Academic and political debates have extensively explored this situation over the intervening years. The impact on the UN revolves around terms like credibility, coherence, and legitimacy (Mazower 2009; Mingst et al. 2022). There were inevitable impacts on trust in the process of multilateral negotiation, international cooperation, and equity within global frameworks. For me, this experience, subsequent conversations, and research inevitably connected directly to the way that sustainable development was being articulated and projected by the UN system. And by a questioning of the extent to which sustainable development was relevant to nation states with their own vested interests and agendas. This understanding also fed into my conceptualisation of the UN and how the organisation's role as a sustainable development champion impacts the concept's ability to be effectively disseminated.

If the UN is ignored in situations of extreme conflict how can it project a concept whose fundamental narrative is one of shared responsibility and interdependence? In response to this question, in part, I have argued elsewhere, and continue to argue here that the UN is representative of a reflexive modernity. This is, it has the ability to reflect and adapt to complex, uncertain, and paradoxical global forces. Moreover, this reflexivity is fundamental for the organisations relevance and legitimacy on the global stage, as well as its ability to successfully negotiate and manage divergent and conflicting needs. Sustainable development has played a central role in this process and in the following discussion, I will elaborate on the way that sustainable development was interpreted during the 57th UNGA. This will then be connected to the current discourse of sustainable development at the international level and subsequently related to local levels of governance in the following chapter.

Building on the previous discussion about the context of the 57th UNGA, the following outlines the approach taken to explore the relationship between sustainable development and a reflexive modernity. The method was based on the degree to which sustainable development was associated with risk, as well as the integration of sustainable development and its effect on the knowledge structures within the UN. The research resulted in four overarching themes that were able to categorise the many and contested ways that sustainable development was understood and embedded within the United National systems. As well as identifying themes and

Table 9.1 Sustainable Development within the UN

Themes	*Outline*
Divergence	Examines the perception and ideological positions of sustainable development within the UN General Assembly
Convergence	Examines the association of sustainable development discourse with border perceptions of risk and suggests a foundation on which sustainable development governance can be based
Dissolution	Examines the way that sustainable development is integrated into resolution negotiations with a particular emphasis on knowledge formation
Evolution	Examines the consequences of sustainable discourse for the UN and more broadly reflexive modernity

events that relate to sustainable development these themes also act as a diagnostic lens for exploring the relation between sustainable development and a reflexive modernity. Table 9.1 presents these themes developed to organise the vast and contested perceptions of sustainable development within the UN.

Table 9.1 presents the four themes and outlines their basic role in understanding the relationship between sustainable development and reflexive modernity. Divergence examines the diversity and ambiguity of perceptions of sustainable development that exists within the 2002 UNGA. Convergence recognises the consistent association of risk with narratives of sustainable development. This convergence of risk represented through sustainable development rhetoric is restructuring governance frameworks at the international scale, which fits the criteria for identifying reflexive modern processes.

Dissolution examines the way that sustainable development affects the certainty of knowledge claims within the context of the United National system and national discourse. Evolution describes the role that sustainable development has played in the restructuring of the UN and the consequences this has for wider society and the emergence of a reflexive modernity. The following discussion will outline their perspectives.

Divergence

The UN is by definition a site of contested ideologies (Knight and Kraus 1995). With this understanding divergence indicates the often fractured and abstracted nature of sustainable development discourse within the UN from a predominantly nation state perspective. Beck (1999, 2006) outlines the changing nature of the nation state in the face of globalised risk, as a prerequisite for identifying a transition from a first industrial to a second/late modernity. Predominantly, perspectives were drawn from the United National General Debate Statement which provided an initial basis for assessing the relationship of sustainable development with a reflexive modernity. By doing so, the divergence and ambivalence of sustainable development discourse within the UN are highlighted. Divergence also refers to the extent to which narratives and discourses of sustainable development represent a departure from, or adherence to the current modes of production and consumption. This was identified as an extension of the narratives that were emanating from the World Conference on Sustainable Development which took place directly prior to the data collection. There was a strong theme of redirection of development and a greening of capitalism that resulted from the conference that created an enabling narrative for countries and organisations across the globe. In this interpretation therefore, sustainable development is not about an identifiable end state, but points to a process (Kemp et al. 2005).

Overall, four main issues, which reflected the political tensions prevalent in the 2002 General Assembly, were very clearly evident in the statements that countries presented. The debate convened on the first anniversary of the September 11th attacks on the World Trade Centre in New York. It was of little surprise therefore that the spectre of global terrorism was high on the political agenda. During the data collection period there were constant references to the horror of that event. This was evident not only within the walls of the United National buildings, but on the streets of New York. The second global priority was the HIV/AIDS pandemic. This disease was emphasised by many of the African nations and was echoed by many more. Thirdly, issues surrounding different aspects of globalisation were discussed. And finally, environmental degradation was targeted variously and in different contexts. Connected to these issues in varying degrees was sustainable development. To organise the breadth of issues discussed, and recognising that these issues are interrelated, a three-pillar analysis was applied to the data, categorising issues and narratives relating to sustainable development as social, economic, or environmental.

Within the economic pillar, issues such as foreign direct investment, globalisation, national development plans, and levels of gross national product were discussed. Similarly, the social pillar emphasised issues of power relations, equality, democracy, health, gender, and terrorism. The environmental pillar is related to such topics as destruction of ecosystems, climate change, sea level rise, desertification, floods, and radioactive waste. As already discussed, these categories are not mutually exclusive and are presented variously within the statements in different ways. But beyond the substantive issues represented by heads of state and national representatives what was witnessed in the general debate was a landscape of ideologies and perspectives, often beyond the scope of categorisation as one blended seamlessly into another. It was also evident that each nation used different discursive tactics to present its definition of what sustainable development should represent, and how it should be achieved, even if these were often vague and unsubstantiated. The result was a kaleidoscope of ideological frameworks and discursive representations of sustainable development. Given the diversity of issues involved as well as the embedded cultural and socio-political perspective of each national context this is inevitable. The UN, as a forum for negotiation, cooperation, and dissent, embodies all the elements and paradoxes of governance structures. This is a fundamental acknowledgment that contradictory goals and various positions exist in political discourse, and that this discourse is impacted by the power structures in the global political system. What was very evident at the time was the broad spectrum of issues contained under a single concept. Perceptions of these issues were not static, they were constantly altering, as certain topics were raised or relegated on the political agenda, as opinion changed, as allegiances altered or, indeed, as the researchers' understanding of the issues evolved.

Ultimately, what was clear was that sustainable development as a concept within UNGA and therefore within policy frameworks the world over is no longer consigned to the narrow parameters of environmentalism. This observation was consistent throughout the research period, whether this was evident in documentation, resolution negotiation, or conversations with national delegates and members of the UN secretariat. As previously mentioned, the proximity of the research period to the World Conference of Sustainable Development undoubtedly influenced the direction and prevalence of sustainable development discourse. The conference's 'Plan of Implementation' represented a substantial alteration from that outlined at the UN Conference on Environment and Development in Rio ten years previously.

Key outcomes are represented in Table 9.2.

We can take this a step further here and explore these outcomes framed through a reflexive modern lens (Table 9.3).

Table 9.2 Key Outcomes

Outcome Area	*Description*
Johannesburg Plan of Implementation	The main negotiated outcome, setting out global priorities for sustainable development. Included targets on poverty eradication, water and sanitation, biodiversity, fisheries, and health, amongst others.
Poverty eradication	Commitment to halve the proportion of people without access to safe drinking water and basic sanitation by 2015, aligned with the MDGs.
Sustainable consumption & production	Emphasis on shifting consumption and production patterns, particularly in developed countries, to reduce environmental degradation.
Natural resource protection	Targets to significantly reduce biodiversity loss by 2010 and restore depleted fish stocks by 2015.
Health & development	Addressed the interconnection between health and environment, focusing on improving access to healthcare and tackling diseases such as HIV/AIDS, malaria, and tuberculosis.
Globalisation & trade	Encouraged greater integration of developing countries into global trade systems whilst safeguarding environmental and social standards.
Means of implementation	Reaffirmed commitments to provide financial support, technology transfer, and capacity-building to developing nations.
Type II partnerships	Over 300 voluntary partnerships launched amongst governments, civil society, and the private sector to implement sustainable development projects. These partnerships were meant to complement formal agreements.
Political declaration	A non-binding statement reaffirming international commitment to sustainable development, multilateralism, and intergenerational equity.
Reaffirmation of Rio principles	Reiterated the foundational principles from the 1992 Earth Summit, including the common but differentiated responsibilities principle and the three pillars of sustainable development.

Table 9.3 Reflexive Framing

Reflexive Theme	*WSSD Outcome Area*	*Reflexive Framing*
Self-confrontation with modernity's failures	Johannesburg Plan of Implementation	Represents an institutionalised moment of self-reflection on development pathways, revealing awareness of unsustainable patterns embedded in industrial modernity and the need for transformative global coordination.
Recognition of manufactured risks	Poverty eradication	Acknowledges that poverty is not a natural state but a product of global systems of inequality. The global response reflects a reflexive awareness of development-induced vulnerabilities and their systemic roots.
Re-evaluation of consumption norms	Sustainable consumption & production	Directly confronts the unsustainability of Western consumption patterns. Calls for behavioural and systemic change signal a shift from first to second modernity, where the logic of unlimited growth is questioned.
Ecological reflexivity	Natural resource protection	The prioritisation of biodiversity and fish stock restoration reflects growing recognition of planetary boundaries and ecological interdependence a core concern of reflexive modernisation.
Health as societal risk nexus	Health & development	Position health crises as manifestations of broader systemic failures where environmental degradation, inequality, and governance gaps intersect, demanding integrated and adaptive responses.
Global governance of risk	Globalisation & trade	Addresses the paradox of global integration producing both opportunities and risks. Promotes regulated globalisation, demonstrating reflexive adaptation in international economic governance.
Institutional reflexivity	Means of implementation	Acknowledges that development initiatives require not only moral intent but structural support highlighting the need for reflexive policy tools like technology transfer and capacity-building to navigate complexity and asymmetry.
Decentralised reflexive governance	Type II partnerships	Signifies a shift towards multi-actor governance models beyond the state. These voluntary, cross-sectoral arrangements reflect the pluralisation of authority characteristic of reflexive modernity.
Normative reflexivity	Political declaration	Symbolic reaffirmation of core values equity, sustainability, and cooperation – suggesting a moral reflexivity that seeks to re-anchor development in ethical imperatives rather than pure economic rationality.
Temporal reflexivity	Reaffirmation of Rio principles	Emphasises intergenerational justice and long-term responsibility, both central tenets of reflexive modernity's concern with the future consequences of present-day action.

The expanded sustainable development narrative that flowed from the WSSD was evident throughout the UNGA where the original focus of sustainable development on the environment was consistently seen as inappropriate in the contemporary political climate.

Aligning sustainable development with only the environment was perceived as not resonating with wider society producing a rhetoric of exclusion which was not conducive with the need to open up the governance process to many different actors within society. Conversations with United National Environment Programme officers confirmed that there was a need to project sustainable development as more than environmental. There was also a recognition that 'the environmental' and sustainable development were being used interchangeably, and following the Rio conference UNEP's mandate was not just environment but also to take into account development. And there was an understanding that broader civic and business organisations needed to be engaged to achieve a sustainable development, or there was a risk that not only organisations but also the concept of sustainable development was being marginalised.

There was a definite aim to distance the concept from its earlier incarnations and make clear that the WSSD represented a turning point in the concept's evolution far beyond its inception at Rio. Whilst the 1992 Earth Summit seems to have indoctrinated the concept into the political agenda, it was seen as being too abstracted from the realities of the political world. With that in mind there was optimism throughout the UN framework that the narrative had moved forward since Rio and the WSSD marked a point where the political apparatus was advancing in a direction that more realistically represented the challenges of the world at that time.

And indeed we are able to witness the progression of this trajectory that eventually became enshrined in the UN Sustainable Development Goals (SDGs) in 2015. What was fundamentally identified was that sustainable development was being utilised by the majority of the world's governments as an indication of good governance, even if the practical goals of achieving this are ill-defined. Also, there was circularity, instead of a linear understanding of the complex environment, and in turn how these environmental phenomena are effecting humanity. What was also very evident was that narratives were associated with varying levels of risk. This is a political discourse which consistently acknowledges the destructive nature of current development patterns and is an initial indication of a reflexive modernity. The nature of this risk was both domestic and global and at this point we explore the notions of convergence.

Convergence

As we have previously identified, reflexive modernity is one where the central institutions of that modernity are forced to react to increased global risk. Observations within the UN focused on major groupings of countries to explore how risk was embedded in these blocks of countries. Whilst the lived realities of these countries are of course very different, the overall association of risk is seen as a consistent theme. Analysis was conducted through the national groupings of the Group of 77, now containing 134 members, the European Union, the Association of Small Island States, the League of Arab Nations, and the Organisation for Economic Cooperation and Development. What was evident was a surprisingly even distribution of perception amongst their national blocks. Linking risk to sustainable development operates as a starting point, methodology to assess the emergence of a reflexive modernity.

As we have discussed throughout the pages of this book in order for sustainable development to represent a reflexive modernity, the risk associated with it must engender a new form of political cooperation that extends beyond the national territory. And within this is what was observed during the 2002 UNGA. The discourses of sustainable development, whilst diverse, point to a fundamental shift in the way nations were perceiving their relationship to internal risk and that of a more globalised risk. Whilst each country articulated sustainable development with reference to their own concerns it was evident that national and global risks were being incorporated into the nationalist discourse. It is suggested therefore that sustainable development represented a globalised risk that encompassed new forms of political cooperation. The internal risk of nations was consistently associated with expressions of wider global risk.

Overwhelmingly, global warming was a threat which was alluded to repeatedly within the context of sustainable development, the complexity of which was easily accommodated with the ambiguous rhetoric of sustainable development. This then inevitably feeds into the broader debates of the relationship between sustainable development, global risk, and the hollowing out of the nation state. For Beck the importance of the altered structure of the nation state in processes facilitates a reflexive governance drawing on multiple stakeholders across sectors.

Convergence illustrates a number of points. Firstly, the important role that risk plays in the discourse of sustainable development. Secondly, it is argued that through the acceptance of the need to develop structures and goals to address these threats new forms of multilateral discourses are becoming evident. Thirdly, the discourse of sustainable development also represented the inclusion of numerous sectors of society through the notion of partnership. Ultimately, it is argued that these observations point to both reflexive governance processes and by extension elements of a reflexive modernity. However, this is only one element of the relationship between a reflexive modernity and sustainable development. The following theme builds on observations and analysis from UN resolution negotiations.

Dissolution

Observing the integration of sustainable development into resolutions, not only from a textual point of view but also from an active negotiation setting enabled a living assessment of sustainable development role in governance processes. And whilst it is recognised that UN resolutions are not legally binding, they are essential for maintaining diplomatic coherence and political capital. This is especially pertinent where negotiations can be witnessed at every stage of the negotiation process. Table 9.4 details the resolution negation phases.

Some themes present themselves immediately to the observer of the multinational negotiation process. For example, the international distribution of power on

Table 9.4 Resolution Negotiation

Stage	*Description*	*Key Actors*
Drafting the resolution	Initial text (zero draft) is written, usually by a sponsoring country or group ('penholder'). May draw from previous resolutions, international agreements, or new priorities.	Sponsoring Member States; informal allies
Informal consultations	The draft is circulated to other delegations. Line-by-line negotiations occur informally. Sensitive terms are debated, and compromise language is introduced. This stage may involve regional blocs or coalitions.	All interested Member States; regional groups
Formal submission	Once there is sufficient support, the draft is formally submitted to the relevant committee or body and given a UN document number. Cosponsorship opens.	Penholders; UN Secretariat
Committee negotiation	Resolution is debated in a main committee (e.g., First, Third, or Fifth). Amendments may be introduced. Some resolutions proceed by consensus; others go to a vote.	Committee Members; all UN Member States
Action on the draft	Committee votes on amendments and the resolution. If passed, the resolution moves to the General Assembly plenary for final adoption (or is directly adopted in the case of other UN bodies, like the Security Council).	Committee Chair; Member States; Voting Delegations
Adoption and publication	The resolution is officially adopted and published in the UN Digital Library. General Assembly resolutions are politically significant but non-binding; Security Council resolutions can be legally binding.	General Assembly or relevant UN body
Follow-up (if applicable)	Some resolutions include mechanisms for follow-up, such as mandated reports, expert panels, or future negotiations. These mechanisms help track implementation or revisit the issue.	UN agencies; special rapporteurs; expert bodies

the behaviour of the states within these negotiating forums. Discourse often represented the preferences of the more powerful actors in the international system and supported observations relating to development discourses as a political strategy of coercion and colonialism (Escobar 2015; Sachs 1992, 1999).

The way that the plan of implementation was interpreted also varied greatly amongst the member states which represented divergent perspectives on how sustainable development should be integrated into governance frameworks. There was also a strong association with the role that science plays in establishing priorities for sustainable development. There was a consistent need for further scientific evidence which is fundamentally a reflexive act within the structure of the resolution negotiations. It is an acceptance of the complex nature of both social and ecological systems, as well as the incomplete nature of current forms of knowledge about the interaction between the two. Where sustainable development was discussed within resolutions, there was often a reference to the need to increase scientific knowledge.

The science of a particular issue was often qualified within the socio-economic parameters, recognising that a unified and universal standard is inappropriate where diverse interests are involved. It was frequently recognised that the implementation of sustainable development would not be achieved through the strict application of scientific standards which adds both a pragmatic approach and complexity to the negotiation process. These observations resonate with discussions in Chapter 2 on the blurring of boundaries between science and policy. Chapter 2 highlighted the importance of the science/policy nexus within the risk society. Within the UN, this interaction directly effects the way that sustainable development is presented and negotiated. There was also significant recognition of the role that scientific and policy uncertainty plays in the narratives of sustainable development. As Voss et al. (2006) point out: 'Sustainable development as a policy goal focuses on long-term dynamics of socio-ecological systems in a global context and transgresses traditional knowledge domains' (2006:8). What was observed within the context of the UN was that in parallel with the earlier recognition that sustainable development is a dialogue that represents the interests of the most powerful nations, the concept also has the ability to dissolve inequalities amongst nations at the discursive scale. This is achieved through its ability to convey to all nations a common understanding of a global community at risk. Understanding sustainable development discourse from this perspective, associated with global risk enables us to draw closer connections with a reflexive modernity. Extending this analysis, the following section explores the way that sustainable development influences the wider UN structure and, in so doing, its consequences for wider governance frameworks.

Evolution

There are two consistent elements identified as evolutionary processes within the UN. The first relates to the changing nature of the perception of sustainable development from an environmental paradigm to one that affects all facets of the social world. We have already seen in this chapter that there were divergent discourses on sustainable development, and that this diversity can be associated with internal and external national risks. This in turn reorders global politics to address these risks pointing to reflexive modern processes. As Kemp (2006) recognises governance is reflexive when it acknowledges that the activities of governance itself are embedded in wider external forces and partly formed by the side effects of its own activities.

With the above in mind we can argue that sustainable development represents a state of reflexive governance on two levels. Firstly, the UN capacity to function as a self-aware governance structure, by championing sustainable development at the global level, and reaffirming its prominent position in global politics. This can be described as purposeful reflexivity and ultimately suggests that sustainable development within the UN creates a language of collective action.

These four overarching themes build a progressive picture of the relationship between sustainable development and a reflexive modernity. The concept is having a profound influence, not only on governance frameworks at the national and supranational level, but also, more fundamentally, facilitates a scrutiny on the interaction between humanity and the environment that effects the very processes of knowledge creation upon which the political processes lie. And based on these observations, it is suggested that sustainable development is representative of a reflexive modernity, supporting Beck's fundamental assertions, with risk infiltrating the central functions of modernity, science, knowledge formation, and political rhetoric.

Moreover, the rapid acceleration of the role of sustainable development on the global stage highlights the importance and relevance of the previous discussion. The establishment of the 17 SDGs has cemented the concept of sustainable development into the future of humanity and planetary health, as such continued interrogation and deconstruction of the context within which the concept is employed is necessary to ensure equitable and consistent application. The following discussion will explore the continued evolution of the concept of sustainable development in the global state.

A Static Shift of International Discourse

My previous observations focused on the discourses emerging from the UN, capturing a moment when sustainable development was still largely framed through A21 and the MDGs. At that time, the rhetoric surrounding sustainable development was characterised by a global emphasis on the need for economic growth to be harmonised with environmental protection and social equity, though inevitably the operationalisation of these principles was uneven and often abstract.

Accepting that there are multiple and contested discourses of sustainable development, there has been a subtle and inconsistent shift in the global rhetoric. The clearest articulation of this shift is evident in the adoption of the 2030 Agenda for Sustainable Development (United Nations 2015), which, as we have already seen, established 17 SDGs as a comprehensive framework for addressing interlinked global challenges. Sustainable development discourse at the international level has further been developed in subsequent reports such as the Global Sustainable Development Report 2023, 'Times of Crisis, Times of Change: Science for Accelerating Transformations to Sustainable Development' (UNSDGs 2023), which emphasises systems thinking, planetary boundaries, and equity. These strategies, policies, and goals suggest a more integrated, intersectional, and transformational language in how sustainable development is conceptualised and promoted at the international level. It is reasonable to conclude therefore that the discourse of sustainable development has shifted both in the way that it is framed, and in the way that it is then articulated and operationalised. The UN' discourse has increasingly emphasised the multidimensional and the 2023 Global Sustainable Development Report reinforces the UN commitment to systems thinking, calling for transformative shifts that span multiple policy domains and sectors.

Also, there are visible and yet to be ascertained implications of the COVID-19 pandemic, but this global phenomenon undoubtedly contributed to a shift by exposing structural vulnerabilities in health systems, economies, and even the very social fabrics of our lives. As highlighted in the UN Secretary-General's series of policy briefs on COVID-19 and recovery (United Nations 2020). These documents call for a sustainable, inclusive recovery, emphasising the need to 'build back better' through policies that simultaneously address public health, environmental sustainability, and social protection. The 2020 UN DESA report also reinforced this perspective by advocating for long-term recovery strategies that align with the SDGs and enhance systemic resilience (UN DESA 2020). And on the point of actual progress towards the goals, the most recent report suggests not only limited progress but in some cases a reversal of key indicators.

But how then do we reconcile the potentially slowing down of sustainable development, as a result of COVID, and the implications this global pandemic has for a reflexive modernity.

Whilst the pandemic slowed and even reversed progress, from a technocratic perspective, it also seems to reinforce Beck's claim that global risks are both destabilising and reflexive. The impact of COVID-19 seems to embody the characteristics of a risk society, where hazards produced and disseminated through late modern interconnected systems create vulnerabilities that permeate national boundaries. As such, these risks demand reflexive engagement, compelling states, institutions, and communities to reconsider their developmental pathways. In this sense, COVID-19 identified a paradox in the relationship between sustainable development and reflexive modernity. Whilst COVID-19 has disrupted concepts of progress, it has also accelerated debates on equity and systemic transformation. Perhaps then instead of understanding COVID-19 simply as a setback for achieving a sustainable development, it can be interpreted as an infection point. An event that identified the fragility of the current system whilst identifying a need to reinterpret sustainable development in more reflexive terms. Extending this analysis further, we can again draw synergies with a risk society, this time through the lens of COVID-19 as an emancipatory catastrophism (Beck 2016). An emancipatory catastrophism suggests that whilst destructive, global crisis can create spaces for opportunity for social and political change.

With the previous statement in mind. it is instructive at this point to engage with relevant critical literatures to build a fuller picture of sustainable development discourse. In light of the significance of the establishment of the SDG, we will begin by trying to understand narratives of sustainable development especially embodied within the goals and targets of the SDGs. We begin with de Vries (2019) who has directly addressed the question as to what extent the SDGs go beyond modernity concluding that '…the SDGs represent the worldview of modernity and are therefore partial and one-sided, with the good life formulated as an abstract enlightenment ideal and targets formulated in a libertarian–utilitarian setting' (de Vries 2019:7).

This argument highlights that a libertarian-utilitarian setting privileges individual agency with minimal structural intervention, and this is reinforced through target setting that is capable of optimising global welfare. This is then representative of the technocratic interpretation of the SDGs and an extension of this to the local scale could indicate a narrative that is also technocratic reducing real transformation. Consequently, this runs the risk of local populations dismissing or rejecting sustainable development '…because of conflicting interests, divergent values and beliefs, ignorance and distrust' (de Vries 2019:9). And indeed, this interpretation of the SDGs reflects my earlier observation of how the concept is transmitted to the local level which can lead to a counter-reflexivity (Borne 2009).

However, it is argued here that whilst de Vries' analysis is important, relevant, and certainly representative of certain dimensions of the relationship between sustainable development and a reflexive modernity, it only represents a partial analysis. What I suggest here is that the evolution of the concept of sustainable development, especially through the medium of the SDGs further reinforces the evidence base for the emergence of reflexive modern processes. It is because of, and not in spite of, sustainable development's inherent ability to draw together conflicting interests, divergent values and beliefs, that is creates the space for real transformation.

With that said we must remain vigilant when interrogating the articulation and operationalisation of sustainable development. In order to achieve this, the following further explores some of the criticisms and pitfalls of sustainable development. As Escobar (2015) points out discourses of development have been identified as perpetuating existing neoliberal ideologies obscuring fundamental and consistently identified structural inequalities in the global system. Reinforcing this position, Bexell and Jönsson (2017) argue that the SDGs risk promoting a post-political consensus that masks global power imbalances. And more fundamentally, the continued discursive institutionalisation of sustainable development where environmental rhetoric has become intimately embedded with global governance but is rendered inert through the removal of real system shifts (Dryzek 2013; Hajer & Versteeg 2005). Moreover, Scoones et al. (2020) suggest that the expansion of terms such as resilience, just transition, and planetary boundaries have serviced to achieve little except dilute even further the ability of sustainable development to effect real change. Scoones et al. also suggest that achieving the SDGs by 2030 will require systemic shifts.

At the policy level, De Jong and Vijge (2021) have analysed storylines relating to the transition from the MDGs to the SDGs in order to explore the evolving narratives. The authors indicate that the MDGs through their founding document the Millennium Declaration '…mainly focused on (economic) poverty reduction without a clear focus on other development dimensions such as human rights, empowerment and equality' (2021:3). These authors continue to develop their arguments suggesting that the way the MDGs and SDGs are initially produced are '…important indicators for the evolution in discourses around sustainable development that find concrete application in global and national governance strategies' (2021:3). A key distinction made is that the MDGs were created with limited consultation using technocratic processes and the SDGs being formulated by multi-stakeholder consultations around the world. Also the MDGs focused on developing countries, whilst the SDGs apply to all countries. And to that end the SDGs are universally applicable. Drawing these observations together we witness both a significant transition in language, narrative, and integrative operationalisation.

Kanie and Biermann (2017) observe a shift from goal-setting to the governance of transitions, emphasising policy coherence, inclusivity, and participatory processes. Here it is suggested that rather than simply extending the modernist project of progress, the SDGs recognise complex, interrelated global risks, and that by their very nature demand, at all levels new forms of critical institutional and self-reflection as well as collective responsibility and effective multilevel governance. From this perspective, the SDGs do not represent de Vries's liberal – utilitarian template, but instead an evolving framework that is able to accommodate diversity, uncertainty, and reflexivity in global development pathways.

At this point in the discussion I pause to recognise that we have come full circle, where we have taken a critical and I hope balanced approach to the positive and negatives of sustainable development discourse. This has been achieved by progressively exploring discourses of sustainable development at the international level, exploring some key policy documents and applying this to relevant critical literatures that shed light on how the global narrative of sustainable development is being framed. From discourses identified through my own observations within the UN to current policies and explorations of the framing of sustainable development I suggest the following speculative typology of sustainable development discourse through key epochs (Table 9.5).

During the Foundational Phase (1987–1992) sustainable development was introduced to mainstream global policy and was influential in shaping early environmental governance with a particular focus on ethics and intra- and intergenerational equity. As we have already seen, LA 21 formed a key policy structure for embedding sustainable development at the local level and particularly within local councils and communities. The Institutionalisation Phase (1992–2015) was characterised by the technical implementation of sustainability through planning, policy, and monitoring sustainability discourses were largely depoliticised, framed in terms of efficiency, partnership, and performance. This is the discourse context for your 2006–2010 town and parish council data, where sustainable development was often equated with practical action, planning constraints, and environmental stewardship. The Transformational Phase (2015–2020) catalysed by the SDGs, which reframed sustainable development as a universal goal rather than an issue for the Global South. This emphasised the integration of environmental, social, and economic systems, highlighting interdependence. Discourse shifted towards values: inclusion, equality, gender rights, and climate justice.

The final stage has been designated the crisis conscious and polycrisis phase (2020–Present). This represents the multiple, interrelated, and parallel issues that are present in the contemporary world. This includes the global pandemic COVID-19, awareness and action in relation to climate change as well as other significant global issues. As we have already seen it is in this space that we glimpse the possibility of an emancipatory catastrophism.

Conclusion

Using this typology, we can position the local discourses of parish councils in Cornwall within the institutionalisation phase of sustainable development discourse. In this phrase, sustainable development is enacted through frameworks

Table 9.5 Shifts in Sustainable Development Narrative

Phase	*Period*	*Global Frameworks/ Events*	*Key Discursive Characteristics*	*Main Narratives*
Foundational Phase	1987–1992	Brundtland Report (1987) Earth Summit/ Agenda 21 (1992)	Introduction of sustainability as balance between growth and environmental protection Strong emphasis on intergenerational equity	Sustainable development as reconciling environment and economy
Institutionalisation phase	1992–2015	Agenda 21, MDGs (2000) Johannesburg Summit (2002) Rio+20 (2012)	Managerial and technocratic State-centric and policy-driven Global North leadership Fragmented across sectors	Sustainable Development as implementation challenge Best practices, capacity building, governance
Transformational Phase (SDG era)	2015–2020	2030 Agenda SDGs (2015)	Universality and integration Systems thinking Framing sustainability as transformation Strong normative emphasis	Leave no one behind Sustainable development as systemic transformation
Crisis-conscious/ polycrisis phase	2020–Present	COVID-19 pandemic COPs 2023 SDG Report	Resilience, vulnerability, and adaptation Post-growth and decolonial perspectives Planetary boundaries and tipping points	Build back better Sustainable development in the age of disruption Justice and equity at the core

such as the MDGs, A21, and projected through events such as the Earth Summit. It is then operationalised through national and regional structures, both formal and informal, as we have discussed in previous chapters. However, as we reflect on the data presented and the issues that are raised in those chapters, we see that in reality, there is a fluidity in the sustainable development discourse that crosses the boundaries between discursive epochs. The following chapter will connect the global analysis to the local dimensions of sustainable development discourse presented in this book.

Bibliography

Beck, U. (1999). *World risk society*. Polity Press.

Beck, U. (2006). *Cosmopolitan vision*. Polity Press.

Beck, U. (2016). *The metamorphosis of the world*. Polity Press.

Bexell, M., & Jönsson, K. (2017). Responsibility and the United Nations' sustainable development goals. *Forum for Development Studies*, *44*(1), 13–29.

Borne, G. J. (2006). Sustainable development: The reflexive governance of risk (Thesis, University of Plymouth). Available at: https://doi.org/10.24382/3702

Borne, G. (2009). Achieving sustainable lifestyles or encouraging a counter-reflexivity: Exploring motivations for sustainability in a mediated risk society. *Local Environment*, *14*(1), 93–107.

de Vries, B. J. (2019). Engaging with the sustainable development goals by going beyond modernity: An ethical evaluation within a worldview framework. *Global Sustainability*, *2*, e18.

De Jong, E., & Vijge, M. J. (2021). From millennium to sustainable development goals: Evolving discourses and their reflection in policy coherence for development. *Earth System Governance*, *7*, 100087.

Dryzek, J. S. (2013). The deliberative democrat's idea of justice. *European Journal of Political Theory*, *12*(4), 329–346.

Escobar, A. (2015). Thinking-feeling with the earth: Territorial struggles and the ontological dimension of the epistemologies of the South. *Revista de Antropología Iberoamericana*, *11*(1), 11–32.

Hajer, M., & Versteeg, W. (2005). A decade of discourse analysis of environmental politics: Achievements, challenges, perspectives. *Journal of Environmental Policy & Planning*, *7*(3), 175–184.

Jabareen, Y. (2004). A knowledge map for describing variegated and conflict domains of sustainable development. *Journal of Environmental Planning and Management*, *47*(4), 623.

Kanie, N., & Biermann, F. (Eds.). (2017). *Governing through goals: Sustainable development goals as governance innovation*. MIT Press.

Kemp, R. (2006). Goal-orientated modulation as a model for dealing with problems of sustainable development. Paper presented at workshop governance for sustainable development, steering in contexts of ambivalence, uncertainty and distributed control, Berlin, February 5th–7th 2006.

Kemp, R., Parto, S., & Gibson, R. B. (2005). Governance for sustainable development: Moving from theory to practice. *International Journal of Sustainable Development*, *8*(1–2), 12–30.

Knight, A., & Kraus, K. (1995). *State, society and the UN system: Changing perspectives on multilateralism*. United Nations University.

Mazower, N. (2009). *No enchanted palace: The end of empire and the ideological origins of the United Nations*. Princeton University Press.

Mingst, K. A., Karns, M. P., & Lyon, A. J. (2022). *The United Nations in the 21st century*. Routledge.

Orosz, A. (2025). *Inside the sustainable development goals: From fighting poverty to saving the world*. Taylor & Francis.

Sachs, W. (1992). *The development dictionary: A guide to knowledge as power*. Zed Books.

Sachs, W. (1999). *Planet dialectics: Explorations in environmental and development*. Zed Books.

Scoones, I., Stirling, A., Abrol, D., Atela, J., Charli-Joseph, L., Eakin, H., … & Yang, L. (2020). Transformations to sustainability: Combining structural, systemic and enabling approaches. *Current Opinion in Environmental Sustainability*, *42*, 65–75.

UN Department of Economic and Social Affairs (UN DESA). (2020). Recover better: Economic and social challenges and opportunities. www.un.org/development/desa/en/news/policy/recover-better.html

United Nations. (2015). *Transforming our world: The 2030 agenda for sustainable development*. United Nations.

United Nations. (2020). UN comprehensive response to COVID-19: Saving lives, protecting societies, recovering better. https://www.un.org/en/coronavirus/UN-response

UNSDGs. (2023). Times of crisis, times of change: Science for accelerating transformations to sustainable development. https://sdgs.un.org/gsdr/gsdr2023

Voss, J.-P., Bauknecht, B., & Kemp, R. (Eds.). (2006). *Reflexive governance for sustainable development*. Edward Elgar.

10 Sustainable Developmental and Reflexive Modernity Connecting the Global to the Local

This final chapter will draw together the key elements of this book, reflecting on the relationship between sustainable development and a reflexive modernity. Initially, I will reintroduce the symbolic relationship between sustainable development and a reflexive modernity that we identified in Chapter 2, which has formed the underlying epistemological and methodological element of research design and subsequent. This is followed by situating the local discourses discussed in earlier chapters within the sharing policy cycles at both the global and local scales. By doing so we see how changing institutional and ideological contexts have shaped and altered the local context of sustainable development discourse. This will be followed by systematically discussing the reflexive indicators outlined in Chapter 4 that have been applied to the local level of analysis. These indicators will then be expanded to incorporate the global level of analysis in order to make key comparisons between the different scales of analysis. This chapter then moves to consider the wider implications of these findings for international discourse and policy. Finally this chapter offers potential avenues for future research.

Initially it is pertinent to revisit the underlying principles that have informed and driven the research presented in this book. This is the symbiotic relationship in the form of synergies that shaped the structure and context of Chapter 2. As presented in Table 2.2, it is fundamentally suggested that the rise in global and local discourse of the concept of sustainable development is representative of a reflexive modernity. This is based on four principal synergies that inexorably link these concepts together. Firstly, both fundamentally challenge the relationship between humanity and the environment. Secondly, both concepts draw into question notions of progress, science as the dominant knowledge process, and rationality. Thirdly, both concepts emphasise the changing nature of governance structures in the face of risk. And fourthly, both concepts must evoke a reflection on the relation between the global and local scales of analysis. The development of the subsequent sustainable reflexivity indicators, both in the previous chapter and as we move forward through this chapter, is built on these underlying synergies. In view of this, the following discussion moves to connect the global and local realms of analysis, addressing the implications for practice, policy, and theoretical development.

DOI: 10.4324/9780429297199-10

Local Reflections

Research presented in this book reveals a complex and textured picture of how sustainable development is understood and enacted at the local level. Town and parish councillors frequently frame sustainable development in terms of immediate, tangible concerns such as land use, affordable housing, flood prevention, community cohesion, the protection of local heritage, and more. These priorities are deeply embedded in place-specific histories and geographies of Cornwall and the Cornish inhabitants. As such, local discourses become inevitably pragmatic, incremental, and context-bound and as we have also seen sustainable development was tied to these local lived realities. Often local discourses balance development pressures with conservation values. Issues such as climate change and global warming, biodiversity, and social inclusion do appear in local agendas, often in response to community activism or regional policy frameworks.

In Chapter 1, we identified the changing nature of the political landscape from the period of research fieldwork until today. It is reasonable to say that the past two decades have profoundly impacted the expectations and capacities, and perceptions of parish councils in relation to sustainable development. During the research period, sustainable development was beginning to emerge in the political state and was being integrated into national policy framework. The Sustainable Communities Act (2007) and other mechanisms sought to embed sustainable development across tiers of government. At the same time, climate change strategies and community agendas encourage local authorities – and indirectly parish councils – to contribute to longer term sustainable development outcomes. Yet, as we saw in Chapter 4, the historically marginal statutory position of parish councils and limited formal powers meant that these expectations were beyond their actual capacity. We also see this presented in the survey data in Chapter 5, which highlighted that many councils expressed rhetorical support for sustainable development, but that there was significant uncertainty about what the concept meant and how it could be operationalised at the parish and community level.

However, the establishment of the Coalition government in 2010 marked a shift in ideology and policy. As we have already discussed, the rhetoric of the Localism Act (2011) contextualised devolution as empowering communities, yet the reality was that austerity measures reduced funding and created a more fragmented local government landscape (Clarke & Cochrane 2013; Pill 2024). This sits firmly within Wills' (2016) analysis which identified the paradox of English local government where greater democratisation is associated with greater centralisation. Drawing on the insights presented in Chapter 6, we see how councillors seemed to struggle with reconciling their pragmatic place-based responsibility with the more abstract sustainable development narratives. Similarly, Chapter 7 demonstrates how councillors exhibited reflexive awareness of the changing and potentially contradictory expectations imposed by central government. What this suggests is that policies of austerity not only impacted on resources but also on political subjectivities. These shifts in policy context, hinted at in Chapter 3, altered the multilevel governance dynamic further compounding the expectation that parish councils achieve more with limited resources.

The arrival of the Keir Starmer Labour government points to a recalibration for future policy direction. As we have also seen in Chapter 3, proposals for Great British Energy reforms to the National Planning Policy Framework and aspirations towards net zero and nature recovery, on the surface at least point to a potentially entering of sustainable development aligned to productivity, growth, and place-based renewal. The findings in Chapter 8 suggest that parish councils' deep embeddedness in local identity and community put them in a central position to address emerging sustainable development, place-based agendas. Identified through the lens of a reflexive modernity, these evolving national cycles illustrate how sustainable development has become a concept of continued and contested governance experimentation.

The previous section has focused on the evolving national landscape, but as it is the central theme of this book, the emphasis here is on the relationship between the local and the global. In the previous chapter we explored the global context of sustainable development outlining the evolving phases of sustainable development discourse. In Chapter 2 we also charted the evolution of sustainable development concept, especially through the medium of key events. In Chapter 2, we charted the trajectory of sustainable development from A21 and the Rio Earth Summit, through to the millennium development goals (MDGs), the World Summit on Sustainable Development, and eventually the sustainable development goals (SDG).

At this point we build on these observations in light of the data and analysis presented at the local level throughout this book. Doing so also allows us to situate the previous discussion on local and national policy cycles within the global scale dynamic. As we saw in Chapter 2, the period of time around the World Summit on Sustainable Development in 2002 marked a consolidation of sustainable development an organising principle in global governance. Moreover, this period also represented my initial engagement in a practical and empirical way with the global stage of sustainable development discourse during the 57th United National General Assembly (UNGA). As we have already discussed, building on the earlier Brundtland legacy, the World Summit on Sustainable Development and the republics established MDGs framed sustainable development within the context of poverty reduction, human development, and the capacity building of the Global South (Redclift 2005; Robert et al. 2005). And whilst these agendas were global in context they exerted a subtle, yet discernible influence within national contexts. The emphasis on measurable targets and partnership-based delivery response with national policies in the UK – especially in relation to the Sustainable Communities Act and National Indicator Frameworks – encouraged local authorities to position their policies within a broader global movement.

The adoption of the SDGs reframed sustainable development as a universal, integrated agenda, drawing together environmental, social, and economic dimensions across all countries and sectors (Kanie & Biermann 2017). Also, the SDGs' emphasis on localisation has explicitly engaged the language of subnational actors identifying multilevel governance as vehicles for translating the goals into local action (OECD 2020). However, as we witnessed through data and discussion in

Chapters 5–8, and whilst the data predates the inception of the SDGs, parish councils' engagement with sustainable development discourse at the time was peripheral and inconsistent. As such, any assumption of global policy that is based on local capacity and willingness to engage is problematic (Guha & Chakrabarti 2019; Mohieldin et al. 2022).

As we identified in Chapter 9, the current discursive epoch of sustainable development discourse can be located within the context of polycrisis. That is an era marked by multiple overlapping disruptions, which include, though are not limited to COVID-19, accelerating climate change and other environmental shifts, increasing inequality, and democratic backsliding (UN DESA 2023; Hickel 2020; Oldekop et al. 2020). And with the Organisation for Economic Co-operation and Development (2022) and the United Nations (2023) reporting that progress on most SDGs has either stalled or reversed, financing gaps and institutional fatigue deepening, substantial challenges exist for local engagement into real-time progress. And ultimately, these dynamics are a reflection of identified criticisms of the concept of sustainable development more broadly as well as the barriers to sustainable development engagement already identified in this book. As we have already identified in the pages of this book, these observations reflect the characteristics of a reflexive modernity (Beck 1999). To elaborate, that is, as risks proliferate and interconnect, so governance systems are channelled into cycles of adaptation and self-confrontation. The result of this dynamic is a global governance landscape that is increasingly fragmented, experimental with discourses filtering down unevenly to national and local levels.

Reengaging with the multilevel governance policy context presented in Chapter 3, it seems that these global policy narratives present a shifting normative framework for local action, that never effectively engages with local populations. This observation is certainly played out through the voices of town and parish councillors in Cornwall. Councillors often negotiated and resisted the implicit expectations of global frameworks, compounded by the realisation that these expectations were not effectively communicated, relevant, or consistent. Perhaps these observations underpin a central observation of the research that sustainable development is not experienced as a linear cascade of policy from the global to the local levels, far from it, instead it is a reflexive, negotiated, and contextually embedded process.

As we move towards 2030 what will be the expected outcomes, the evaluation of the progress of the SDGs. For me this is not the culmination of 15 years of the SDGs; instead, as it was with the MDGs it will be a time for reflection and recalibration. A genuine opportunity to reframe global ambitions through the medium of grounded local realities. Ultimately, the success or failure of future global agendas will depend upon their capacity to resonate with and empower local communities.

The following discussion will apply these observations directly to the reflexivity indicators. Recognising that these indicators are not definitive with regard to the boundaries of each category, they do provide a framework for understanding the relationship between reflexive modernity and sustainable development at the local level, I will elaborate more critically on this in the subsequent discussion. At this stage we will, with broad strokes, reflect on how each indicator applies to the findings and analysis presented in the previous chapters (Table 10.1).

Table 10.1 Reflexive Indicators

Indicator	*Local Discourse Signals*
Displacement of traditional authority	Local partnerships, bottom-up decision-making
Knowledge pluralism	Identification of lay/local knowledge, critique of top-down policies, challenging scientific hegemony
Ethical and moral reasoning	Moral imperatives, future generations, justice
Reflexive consumption/choice-making	Local sourcing, sufficiency, conscious practice, ethical decision making, transport choices
Adaptive governance	Experimentation, learning, flexibility
Global-local risk awareness	Links to climate change, sustainable development, local risk in a global context

Firstly, displacement and traditional authority are evident through decision-making that is not represented by top-down processes. Instead, decisions are shaped by horizontal community-led initiatives. This is evident in town and parish councils where we witness collaborative partnerships with community groups, businesses, local communities, and other organisations. To be clear, I am not suggesting that there is an absence of traditional authority but instead a refocus of that authority. As such, the use of deliberative and participatory processes for the creation of sustainable development initiatives is representative of this indicator. Also, this indicator is seen to represent resistance to and even rejection of imposed solutions in favour of locally determined approaches.

The second indicator, knowledge pluralism and the critique of expert systems, recognises the limits of scientific and institutional knowledge and emphasises the inclusion of local, lay, indigenous, and experiential knowledge. This was witnessed throughout the research with multiple statements that value local knowledge, heritage and generational knowledge, and a scepticism of expert knowledge. This indicator also reflects concerns and critical reflection, and even discomfort with centralised policy approaches. This indicator also relates to the inclusion of diverse perspectives in the decision-making process.

The third indicator of ethical and moral reasoning relates to how sustainable development and associated themes are framed in terms of values and responsibilities. This includes intra- and intergenerational equity and justice and moves beyond economic and regulatory narratives. This manifest itself within the research in multiple ways and included language that pointed to notions of duty, responsibility, and stewardship. As previously mentioned, this was also connected to references to future, and past generations evoking sentiments of fairness and moral imperatives. This indicator is also represented by the central focus of ethical considerations in decision-making processes that extend beyond legal and regulatory requirements.

The fourth indicator, sustainable consumption and reflexive choice making, represents narratives that highlight deliberate and reflexive changes to practices and consumption patterns in the face of risks associated with sustainable development. Discourses that represent this within the research are evident in narratives of issue relating to local and green procurement, locally produced food, as well as more localised forms of energy consumption. This feeds into broader narratives of shifts towards individual and community sufficiency. And these narratives are part of a larger critique of consumerism and the logic of growth and imposed progress.

The fifth indicator, adaptive governance, relates to the ability of governance structure to be reflective in the face of sustainable development and related risks. Adaptation related to governance frameworks can be experimental and learning based as well as being responsive and flexible to both internal and external factors. Within the data this was evidenced through expressions relating to learning from past mistakes and adapting strategies. An ability and willingness to experiment with governance structures and an acceptance that structures will inevitably change over time also play a role within this indicator. Also, adaptable governance draws on the language of, and accepting that, uncertainty is an inevitable and unavoidable part of the process. This represents a move away from notions of management and control.

And finally, the sixth indicator identified here is the recognition of the relationship between global and local risk and action that are associated with that risk. This has emerged through explicit connections between local actions related to sustainable development and global risks. For town and parish councils this is visible through the framing of local initiatives that address global responsibility as well as an awareness of local decisions that have global consequences. This is perhaps an inevitable theme within the research as the connection between the global and local is explicit throughout the research design and forms a central component within the survey.

As I have identified already, in this and previous chapters, these indicators were weaved into the very fabric of the research design which provided the bedrock for the initial survey design and well as subsequently being applied to the discourses identified within the interviews. And as already stressed, these indicators are not meant to, nor can they be, applied in a quantifiable manner. Instead, the indicators provide a discursive and metaphorical checklist for exploring the degree to which in-depth interview discourse and responses to the survey data could point to a reflexive modernity, and to that end they have provided effective.

Building on this, it is the goal of this final chapter to synthesise the local context of the book with the global context that we explored in Chapter 9. In order to take that forward we now compare globally cognate reflexivity indicators so that both global and local scales of analysis can be incorporated in the analysis. Table 10.2 connects the global and local reflexive indicators to these reflexive domains.

With the reflexivity indicators applied to both the global and the local levels of analysis, we can now build a more robust picture of the relationship between sustainable development and a reflexive modernity. I have suggested elsewhere that narratives of sustainable development at the global level represented a reflexive modernity whilst at the local level of analysis the relationship is more convoluted,

Table 10.2 Global and Local Reflexive Indicators

Global Reflexive Elements	*Local Reflexive Indicators*	*Description*
Institutional self-reflection Global recognition of systemic unsustainability and need for governance reform	*Displacement of traditional authority* Local partnership, bottom-up governance	Global institutional reflexivity aligns locally with decentralised, inclusive governance practices reflecting shared awareness of traditional governance inadequacies.
Awareness of manufactured risks Understanding that contemporary risks are products of global systemic processes	*Global-local risk awareness* Local contextual understanding of global risks	Both global and local levels demonstrate reflexive awareness that risks and vulnerabilities are constructed by social, economic, and political systems.
Re-evaluation of consumption and growth norms Questioning of unlimited growth and consumption patterns	*Reflexive consumption/ choice-making* Local ethical and informed consumer decisions	Global re-evaluation of unsustainable consumption aligns with local practices of ethical, responsible, and informed consumption choices.
Ecological and knowledge reflexivity Recognition of ecological boundaries and diverse forms of knowledge	*Knowledge pluralism* Incorporation of diverse local and lay knowledges	Global acknowledgment of ecological limits and knowledge diversity is mirrored locally by integrating diverse, pluralistic forms of knowledge into governance and decision-making processes.
Ethical and temporal responsibility Emphasis on intergenerational equity and ethical responsibility	*Ethical and moral reasoning* Local discourse on equity, future-focused sustainability	Global ethical and temporal reflexivity strongly connects to local ethical discourses emphasising intergenerational justice, sustainability, and moral responsibility.
Adaptive and collaborative governance Flexible, multilevel partnerships and policy structures	*Adaptive governance* Local adaptability, innovation, and collaboration	Global movement towards adaptive and collaborative governance resonates locally with flexible, experimental, and multi-stakeholder governance approaches.

and instead of a reflexive modernity, narratives of sustainable development are representative of a counter-reflexivity (Borne 2009).

Concisely, a counter-reflexivity suggests that individuals consciously choose to disassociate from the messaging that was connected to sustainable development. This was occurring for two straightforward reasons. Firstly, because global risks were being inhibited by local lived experience, and secondly, because the messaging of sustainable development lacked salience and relevance to people and communities not directly impacted by risks that were identified as being associated with sustainable development. Conversely, at the global level, sustainable development was seen to represent a reflexive modernity as global institutions and governance processes sought to reconfigure in the face of risk. This dichotomy between reflexive and counter-reflexive processes was not identified through a strict division of knowledge formations between rationalist and expert at the global level and lay and experiential at the local level. Moreover it was suggested that it is the very proliferation of assemblage dynamics that we identify in Chapter 3 that ultimately leads to a rejection and dissociation of the images of global risk.

This analysis, whilst providing important initial insights, is, of course, and indeed necessarily reductionist. Ultimately, this polemic heuristic provides a way of navigating the complexity of global dynamics and extends an invitation to delve more deeply and more comprehensively into the issues it reveals. Accordingly, at the global level, in Chapter 9 I built on this foundation to provide a more sophisticated picture of sustainable development discourse. This was achieved in a number of ways, reframing key policy documents, the Johannesburg Plan of Implementation for example, charting shifts in global sustainable development narratives from the Brundtland report through to the SDGs and critically exploring the construction of transition to the SDGs from the MDGs. Similarly, throughout the previous chapters, the local level data and analysis have built on the initial observations pointing to a far more sophisticated interpretation of reflexive processes than previous observations permitted.

Theoretically, what the data presents us with is a rich, dynamic, and changing landscape of sub-political formation. This allows us to reimage the binary distinctions of reflexivity and counter-reflexivity. The following discussion therefore identifies an expanded conceptual framework for reflexivity grounded in the observations presented in previous chapters. Fundamentally, what is recognised is that the strict separation between reflexivity and counter-reflexivity, especially in the context of identified assemblages and sub-political dynamic space, needs to be treated with caution and explored in more detail (Table 10.3).

Taken together, these expanded categories of reflexivity provide a way of acknowledging that sustainable development discourses operate through a spectrum that ranges from acknowledging deeply transformative processes through to resistance, denial, and broader counter-reflexive processes that operate in the face of globalised risk. Moreover, these reflexive positions demonstrate how local and global narratives are intimately interwoven, yet seldom aligned. Whilst we can suggest that global institutions are capable of embracing reflexive framings, in principle at least, local actors often encounter these framings through the pragmatic constraints of governance processes, or disengage from them entirely. Also, using a spectrum of reflexive positions allows us to move past binary assessments of success or failure and instead see reflexivity as an uneven, contested, and dynamic process. In light of the previous assessment, we can reconnect with the central

Table 10.3 Reflexivity Expanded

Type of Reflexivity	*Description*	*Level(s) Observed*	*Example Indicators*
Deep reflexivity	Critical engagement with the root causes of risk and unstable processes. Willingness to challenge dominant models of development and propose alternatives.	Global UNSDGs-Local	Use of language relating to limits to growth, intergenerational justice, planetary boundaries, systemic transformation. Rejection of scientific hegemony.
Strategic reflexivity	Aware of global risks but uses this awareness to advance self-interest or institutional legitimacy, often without meaningful change.	Governments, corporations, NGOs	SDG branding, ESG reports with limited substantive change, rhetorical alignment without deep structural shift.
Pragmatic reflexivity	Reflexivity embedded in everyday problem-solving and local governance. Actions respond to risk but are constrained by capacity, resources, or competing priorities.	Town and parish councils	Discussions around flooding, housing, land use, and biodiversity – often framed in practical, context-specific ways.
Latent Reflexivity	Potential for reflexivity exists but is unrealised due to limited access to knowledge, institutional inertia, or risk displacement.	Local Individual/ community level	Low engagement with SDGs or climate discourse; prioritisation of immediate risks, cost of living, transport, jobs.
Counter-reflexivity	Risk awareness leads to avoidance, denial, or reinforcement of unsustainable practices. Often results in surface-level compliance or rejection of reflexive discourse.	Global climate denial, greenwashing, local economic self-interest	Energy-saving for cost reasons only; rejection of climate science; apathy towards long-term planning.

proposition of this book that sustainable development is achieved not through a linear policy trajectory but as a manifestation of a reflexive modernity. That is continually reshaped through cycles of critique and adaptation at the global and local levels of analysis.

Lessons for Discourse and Policy

The analysis presented here suggests a number of implications for sustainable development governance. Initially, it is almost a common-sense statement to say that implementation of the SDGs must be embedded at the local level. But this means not simply employing a tokenistic consultation. Instead, it requires a more fundamental configuration of the process one that is rooted in genuine coproduction between diverse stakeholders, not to mention the recognition of multiple systems of knowledge creation. These understandings feed into a recognition that local governance structures, such as town and parish councils, and other configurations across the world hold considerable potential to embed the aspirations of sustainable development.

But as we have discussed, they also face considerable structural and cultural barriers which global policy discourse seems ill-equipped to engage with. Applying the reflexivity typology evidence identified in this book shows that councillors expressed ethical concern, risk awareness, and sense of long-term responsibility, which aligns with a deep reflexivity. Yet actual practices are ultimately shaped by a pragmatic reflexivity. That manifests in the form of themes we have consistently identified, including immediate community needs, limited resources, and institutional constraints. In some cases, we were also able to witness latent and even counter-reflexive tendencies as councils disengaged with global narratives that were perceived as irrelevant at the local level.

The SDGs are a valuable global framework, but this framework depends in a very significant amount on its ability to engage with and learn from the local level. This final chapter has attempted to draw out the parallels, divergences, and opportunities between local and global discourses of sustainable development. By listening to the voices of people that work at the most local levels, we gain a richer understanding of how sustainable development is lived, negotiated, and transformed in everyday governance. By doing so, we set a challenge to the international community to reimagine sustainable development not as a top-down directive, but as a genuinely collaborative and constructed solution.

Future Research

Moving forward, the research and analysis presented in this book open up a number of important avenues for future investigation. Initially, it is accepted that there could be a re-evaluation of the existing data from both the survey and the interviews presented in this book. Results presented here have been presented through a particular analytical lens to achieve the ultimate goal of exploring the relationship between sustainable development and reflexive modernity. But the richness of the qualitative data presents multiple issues that are far beyond the scope of this text to discuss. There is therefore scope to reinterrogate this data both by elaborating on issues already identified and by applying different analytical and theoretical frames to uncover further issues and connections. Similarly, the survey results discussed here have, for practical purposes, focused on descriptive statistics showing frequencies of responses to achieve a particular research goal. Whilst this was what was required to achieve these goals, an analysis could be conducted that explores the relationship between the different issues and variables identified in the survey.

Beyond the existing data, there is significant scope to revisit town and parish councils and local governance structures more generally to explore contemporary discourses of sustainable development. Because the original fieldwork pre-dated the adoption of the SDGs, future work could explore how discourses at the local level have evolved in the face of new global agendas as well as the current global risk priorities. Research moving in this direction could seek to understand how awareness, engagement, and practices have shifted over time. Future work could compare differences between local settings, as well as expanding the research to engage with urban and rural environments and understanding evolving dynamics between different tiers of government and other organisations.

Pushing the boundary of enquiry even further, there is an opportunity to expand this research across national boundaries. For example, applying the reflexive typology presented here to the local context of other countries would provide fertile ground for cross-national comparisons of local governance contexts. Research would be able to explore how sustainable development is enacted under different institutional and cultural conditions. Moreover, internationally comparative studies could further understand how multilevel governance processes are translated, negotiated, and resisted from the global to the local levels, and vice versa.

Moving from international comparisons of local dynamics and reconnecting with the global scale of analysis, it would be valuable to further explore discourses of sustainable development from within the United Nations. A continued and expanded analysis of key documentation, as presented in Chapter 9, as well as further analysis of key documentation could prove fruitful. This could include an exploration of narrative presented during subsequent United National General Debate Statements which are now freely available through the United Nations library.

Theoretically, there is always scope to continue the exploration of the relationship between sustainable development and reflexive modernity. This is an ongoing and extensive agenda and could progress on multiple fronts, too numerous to discuss here. However, further research could refine, refute, or amend the relationship between reflexivity and counter-reflexivity that has been presented in this book. Theoretical refinements could be linked to risk governance deepening our understanding of how sustainable development discourse adapts under conditions of crisis and uncertainty. This is not an exhaustive inventory of potential research directions, far from it, and even as I write this passage topics relating to longitudinal tracking of local discourse, policy ethnographies, comparative analysis of knowledge systems, and more spring to mind. It is my hope that this book provides a platform for multiple and of course as yet unforeseen avenue of future work that ultimately help bridge the persistent gap between global ambition and local reality.

And finally, at least for now, as I reflect on the journey from the time spent at the United Nations in 2002 to the present day, one thing becomes clear. Sustainable development as a concept, policy goal, and practical guide for the future, is here to stay. It now occupies a space with a generation of humanity who have not known a world without it, who are learning about the key concepts in primary and secondary school, as are my own children, to a higher education system that explicitly asks for the SDGs to be acknowledged throughout degree program structures. Yet as we have seen, progress towards the goals is slow, and there is still a great deal of work to be done. Sustainable development is not static; it is contested, uncertain, and ambiguous, but these are exactly the qualities that mean the concept can form a platform for multiple countries, ideologies, religions, and perspectives to come together for a better future.

Bibliography

Beck, U. (1992). *Risk society: Towards a new modernity*. Sage.

Beck, U. (1999). *World risk society*. Polity Press.

Borne, G. (2009). Achieving sustainable lifestyles or encouraging a counter-reflexivity: exploring motivations for sustainability in a mediated risk society. *Local Environment*, *14*(1), 93–107.

Clarke, J., & Cochrane, A. (2013). Geographies and politics of localism: The localism of the United Kingdom. *Political Geography*, *34*, 10–23.

Department for Communities and Local Government. (2008). *National indicators for local authorities and local authority partnerships: Handbook of definitions*. Department for Communities and Local Government.

Guha, J., & Chakrabarti, B. (2019). Achieving the sustainable development goals (SDGs) through decentralisation and the role of local governments: A systematic review. *Commonwealth Journal of Local Governance*, *22*, 1–21.

Hickel, J. (2020). The sustainable development index: Measuring the ecological efficiency of human development in the anthropocene. *Ecological Economics*, *167*, 106331.

HM Government. (2007). *Sustainable communities act 2007*. The Stationery Office.

HM Government. (2011). *Localism act 2011*. The Stationery Office.

HM Government. (2024). *National planning policy framework: Draft update*. Department for Levelling Up, Housing and Communities.

Kanie, N., & Biermann, F. (Eds.). (2017). *Governing through goals: Sustainable development goals as governance innovation*. MIT Press.

Labour Party. (2024). *Labour Party manifesto 2024: Change*. The Labour Party.

Robert, K. W., Parris, T. M., & Leiserowitz, A. A. (2005). What is sustainable development? Goals, indicators, values, and practice. *Environment: Science and Policy for Sustainable Development*, *47*(3), 8–21.

Mohieldin, M., Wahba, S., Gonzalez-Perez, M. A., & Shehata, M. (2022). The role of local and regional governments in the SDGs: The localization agenda. In M. A. Gonzalez-Perez, L. Leonard, & J. C. McDermott (Eds.), *Business, government and the SDGs: The role of public-private engagement in building a sustainable future* (pp. 105–137). Springer International Publishing.

Oldekop, J. A., Horner, R., Hulme, D., Adhikari, R., Agarwal, B., Alford, M., … & Zhang, Y. F. (2020). COVID-19 and the case for global development. *World Development*, *134*, 105044.

Organisation for Economic Co-operation and Development. (2022). *Global outlook on financing for sustainable development 2023: No sustainability without equity*. OECD Publishing. www.oecd.org/en/publications/global-outlook-on-financing-for-sustainable-development-2023_fcbe6ce9-en.html

Organisation for Economic Co-operation and Development. (2020). *A territorial approach to the Sustainable Development Goals: Synthesis report*. OECD Publishing. https://www.oecd.org/en/publications/a-territorial-approach-to-the-sustainable-development-goals_e86fa715-en.html

Pill, M. (2024). Rethinking the 'local state' and local capacity. *Local Government Studies*, *50*(6), 998–1007.

Redclift, M. (2005). Sustainable development (1987–2005): An oxymoron comes of age. *Sustainable Development*, *13*(4), 212–227.

UK Department for Energy Security and Net Zero. (2024). *Great British energy: Policy prospectus*. Department for Energy Security and Net Zero.

United Nations, Department of Economic and Social Affairs (UN DESA). (2023). *Progress towards the sustainable development goals: Towards a rescue plan for people and planet*. United Nations.

United Nations, Department of Economic and Social Affairs. (2023). *The Sustainable Development Goals report 2023: Special edition*. https://unstats.un.org/sdgs/report/2023/.

Wills, J. (2016). Emerging geographies of English localism: The case of neighbourhood planning. *Political Geography*, *53*, 43–53.

Appendix 1
Questionnaire

Understanding Town and Parish Council Needs for a Sustainable Cornwall

Section One: You and Your Parish

Please indicate what parish council area you represent?

Do you consider your parish council area to be…?

☐ Rural ☐ Urban ☐ Mixed ☐ Other ☐ Don't Know

What is your main motivation for being a parish councillor/clerk/other?

☐ Improve your community ☐ Other

☐ Paid local employment ☐ Interest in politics

☐ Increased social standing ☐ Don't Know

If other please indicate:

Do you reside in your parish council area?

☐ Yes ☐ No ☐ Don't Know

If yes, please indicate how long you been a resident in your parish? (Approximately in years)

☐ 0–1 ☐ 11–15

☐ 1–5 ☐ 16–20

☐ 6–10 ☐ 20+

What is your role in the parish council?

☐ Chair ☐ Councillor ☐ Other

☐ Clerk ☐ Employee

Before becoming a parish councillor/clerk/employee/other, were you aware of what the role entailed? (please circle)

Very 1 2 3 4 5 Not at all

What barriers do you think there are to becoming a councillor/clerk/employee / other?

☐ Work commitments
☐ Negative perception of parish council
☐ Lack of understanding of role
☐ Finance
☐ Time
☐ Other

If you answered other to the previous question, could you please comment below:

In your role as a parish councillor/clerk/employee/other, do you think you have had a positive effect on your local community? (Please circle)

Very 1 2 3 4 5 Not at all

How do you think parish councils should make people more aware of their activities?
Will you continue to be a parish councillor/clerk/employee/other in the future?

☐ Yes ☐ No ☐ Don't Know

Section Two: Identifying Parish Council Needs

Do you have a parish plan?

☐ Yes ☐ No ☐ Don't Know

If you answered yes to Question 14, do you feel your contribution to the parish plan could be improved through personal education and training?

☐ Yes ☐ No ☐ Don't Know

Do you feel that your parish council supports community organisations in the parish planning process?

☐ Yes ☐ No ☐ Don't Know

Do you think that your parish plan takes into account the strategic goals of principal authorities?

☐ Yes ☐ No ☐ Don't Know

If you answered yes, please comment below.

If you answered no to the previous question, do you think your parish plan should have some links with the strategic goals and aims of principal authorities?

□ Yes □ No □ Don't Know

Does the strategic planning of principal authorities sufficiently consider local needs?

□ Yes □ No □ Don't Know

Do you think your county council is doing enough to help you in your role as a parish councillor?

□ Yes □ No □ Don't Know

If you answered no to the previous question could you briefly indicate in what ways you think your County Council could help in your role as a parish councillor? Please comment below?

Do you feel it is important to work in partnership to improve the well- being of your community?

□ Yes □ No □ Don't Know

If yes, do you consider any of the following to be important for working in partnership? (Tick where appropriate)

□ National Government □ Voluntary Sector

□ Private Sector □ Don't Know

□ Other □ Other Parishes

□ County Council □ Unitary Authority

Are you currently working in partnership with any of the following to improve the well- being of your community? (Tick where appropriate)

□ National Government □ Voluntary Sector

□ Private Sector □ Don't Know

□ Other □ Other Parishes

□ County Council □ Unitary Authority

Please indicate to what degree you feel the following statements may provide barriers to the provision of adequate services in Cornwall: (please circle)

The scattered layout of Cornwall's Communities:

Very significant barrier 1 2 3 4 5 Not at all significant barrier

Lack of involvement of community and individuals in the planning, development and delivery of services:

Very significant barrier 1 2 3 4 5 Not at all significant barrier

An increasingly ageing population:

Very significant barrier 1 2 3 4 5 Not at all significant barrier

Services not meeting complex needs:

Very significant barrier 1 2 3 4 5 Not at all significant barrier

Do you feel that there are any other significant barriers to service provision not outlined in the previous discussion? Please comment below

Do you feel your parish precept should be?

☐ Increased ☐ Stay the same

☐ Decreased ☐ Don't know

Are you aware of any of the following?

☐ Local Areas Agreements ☐ Quality Parish/Town Council Scheme

☐ Comprehensive Area Assessments ☐ Cornwall's Sustainable Community Strategy

Is your parish involved in any of the following?

☐ Local Areas Agreements ☐ Quality Parish/Town Council Scheme

☐ Comprehensive Area Assessments ☐ Cornwall's Sustainable Community Strategy

Do you feel your parish council meetings are generally effective in addressing relevant issues for your parish? (please circle)

Very 1 2 3 4 5 Not at all

If you feel your parish council meetings are not effective could you briefly indicate possible reasons for this below:

Do you feel your council meetings could be improved in any way?

☐ Yes ☐ No ☐ Don't Know

If you answered yes to the previous question, please indicate in what ways you feel your council meetings could be improved:

Do you feel that your parish council engages with the wider community and promotes participation outside of local organised groups?

□ Yes □ No □ Don't Know

Understanding a Sustainable Community

Do you feel that you are part of a community?

□ Yes □ No □ Don't Know

Do you think you are a member of more than one community?

□ Yes □ No □ Don't Know

Which of the following is your community focused around? (Please tick those that apply)

□ Work □ Family and Friends □ The Internet

□ Your Local Area □ Other □ Leisure Activities

Do you consider your community to be: (Please tick those that apply)

□ Within your local area □ Not confined to a particular area

□ Nationwide □ Don't Know

Please indicate which of the following areas you believe a sustainable community refers to: (please tick where appropriate)

□ Economy □ Environment

□ Society □ Other

If you have any further thoughts on a sustainable community, please indicate below:

Please rank the following in order of importance to you: (please number – 1 being the most important and 6 the least important)

□ Crime □ Transport □ Education

□ Social Welfare □ Environment

□ Health □ Don't know

If other, please briefly comment:

Please rank the following issues in order of concern for Cornwall: (Please number: 1 = highest concern; 8 = lowest concern)

☐ Air pollution ☐ Global Warming

☐ Waste Disposal ☐ Increasing Population

☐ Global Economy ☐ Traffic speed

☐ Water Pollution ☐ Flooding

Now please rank the same issues in order of concern for the whole world. (please number: 1 = highest concern; 8 = lowest concern)

☐ Air pollution ☐ Global Warming

☐ Waste Disposal ☐ Increasing Population

☐ Global Economy ☐ Traffic speed

☐ Water pollution ☐ Flooding

Please indicate which of the following areas you feel are most important for Cornwall: (please tick those that apply)

☐ Economy ☐ Society

☐ Environment ☐ Don't Know

What do you understand as 'the environment'? (Please tick those that apply)

☐ The countryside ☐ Don't Know

☐ Everything on planet earth ☐ The world around you

☐ Nature ☐ Everything outside of society

In your everyday life how important are environmental issues to you? (Please circle)

Very 1 2 3 4 5 Not at all

Do you consider yourself to be environmentally aware? (Please circle)

Very 1 2 3 4 5 Not at all

Do you feel you behave in an environmentally friendly way? (Please circle)

Very 1 2 3 4 5 Not at all

Which of these phrases have you heard of?

☐ Climate Change ☐ Biodiversity

☐ Ozone depletion
☐ Global Warming
☐ Agenda 21
☐ Carbon footprint

If you are aware of the term carbon footprint, please indicate if you are concerned about your parish's carbon footprint?

☐ Yes ☐ No ☐ Don't Know

Do you think global warming is…? (Please tick one)

☐ Not really happening?
☐ Caused by natural processes with nothing to do with humans?
☐ Caused by natural processes but made worse by human beings?
☐ Completely the human race's fault?

If you indicated that you are aware of global warming in Question 65, do you feel Global Warming is an important issue for Cornwall?

☐ Yes ☐ No ☐ Don't know

If yes to the previous question, which of the following do you consider to be the most significant drivers for climate change in Cornwall? (please circle)

Housing stock with poor environmental performance:
Very significant 1 2 3 4 5 Not at all significant

Absence of affordable housing driving pressure for house building:
Very significant 1 2 3 4 5 Not at all significant

Building structures located on flood plains:
Very significant 1 2 3 4 5 Not at all significant

Dispersed settlement patterns no longer supporting self-sustaining communities leading to increased travel:
Very significant 1 2 3 4 5 Not at all significant

Individual ambitions for car ownership:
Very significant 1 2 3 4 5 Not at all significant

Insufficient appraisal and analysis of greenhouse gas emissions:
Very significant 1 2 3 4 5 Not at all significant

Do you think recycling is important for Cornwall?

☐ Yes ☐ No ☐ Don't Know

Have you heard of the term sustainable development?

☐ Yes ☐ No ☐ Don't know

If yes, which of the following areas do you think sustainable development refers to? (Please tick those that apply)

☐ The Environment
☐ The Economy
☐ Society
☐ Other
☐ Don't Know

If other please comment below:

If you answered yes to the previous question, please indicate on the statements below your interpretations of sustainable development (SD): (Please circle)

SD means environmental issues are more important than money:

Agree 1 2 3 4 5 Disagree

SD is the protection of the planet at all costs:

Agree 1 2 3 4 5 Disagree

SD is jargon used by governments to make themselves sound greener:

Agree 1 2 3 4 5 Disagree

SD is a way of preserving natural resources so that they will be available for future generations:

Agree 1 2 3 4 5 Disagree

If you have any further comments on sustainable development please outline these below:

About You

☐ Male ☐ Female

Please indicate your age:

18–25 26–32 33–40 41–48 49–56 57–64 65–72 72+

How many people are in your household? (Please Circle)

1 2 3 4 5+

Are You?

- ☐ Tradesman/woman
- ☐ Student
- ☐ Professional
- ☐ Retired
- ☐ Manual Worker
- ☐ Housewife/husband
- ☐ Unemployed
- ☐ Clerical/Administrative

Is your home?

- ☐ Owner occupied
- ☐ Rented – housing association
- ☐ Rented – privately

What is your highest educational qualification?

- ☐ O Level/CSE/GCSE
- ☐ Post-Graduate
- ☐ A level or GNVQ
- ☐ No conventional qualifications
- ☐ Degree
- ☐ Other (please indicate here)

Index

Note: **Bold** page numbers refer to tables and *italic* page numbers refer to figures.

affordable housing 84, *84*, 98, 105–106, 113, 130, 165
Agenda 21 (A21) 17–20, **18**, **19**, 81, 161, 166; comparison between SDGs and **18**, 18–19; WSSD 18–19
Allen, C. 15
attitudes, behavioural, choice model (ABC) 126

barriers to service provision 65–69
Beck, U. 8, 28, 82, 110; fundamental assertions 157; Risk Society Thesis 1, 11, 23; *Risk Society; Towards a New Modernity* 11; sub-political space 28; *World Risk Society* 11
belonging 6, 95, 130, 138; displacement of 143; second homes tourism 143
Bexell, M. 159
Biermann, F. 159
Biodiversity Convention 17
Brundtland Commission 16; *see also* World Commission on Environment and Development
Brundtland report 166, 171

capitalist market system 12
changing nature of governance structures 14–15
climate adaptation 37, 38, 40–41
Climate and Nature Bill 39
climate change 5, 15, 17, 38, 40, 81–87, 109; in Cornwall 82–87; Framework Convention on Climate Change 17; strategies 165
Climate Change Act 2008 35, 37, 39
Climate Emergency Action Plan 41
climate-focused policy interventions 2
climate governance: local 41; role of local councils in 38–39; strengthening 41
coastal: community 140–141; identity 140–141; isolation 141–142
community: academic 11; action 116; activism 165; actors 36; agendas 165; asset 136–137; -based adaptation measures 37; capacity 30; centres 65; coastal 140–141; coherence 65; cohesion 42, 94, 97, 106, 116, 135–136, 139, 140, 142, 144–145, 165; dynamics 94, 112, *130*; empowerment 36; energy schemes 37; engagement 41–42, 131; global 16; governance 107, 138; groups 168; identity 58, 138, 139, 143; improvement 127; -managed green spaces 31; needs 122, 173; priorities 97, 107; resilience 49, 94, 95, 106; responsibility 116, 132; right 36; sufficiency 169; sustainability 138; unpacked 129–134; wealth-building strategies 42
Community Area Agreements 79
Community Area Partnerships (CAPs) 49, **49**
community-led renewable energy: initiatives 35; projects 27, 37, 39, 40
community-led sustainability 2
Community Network Areas (CNAs) 49, **49**
community-owned renewable energy 127
community spirit 131–132, 145
community stewardship 144–145
comparative perspectives on local sustainability governance 38
complex interacting systems 14
constructivism 13–14
convergence 153–154

Copus, C. 32
cornish context 33–35
Cornwall **48**, 48–49; community-owned/locally invested wind projects in 118, **118**; and global warming 82–87, *83–86*; institutional and community contexts of 51; key statistics **48**; locally invested wind projects in **118**; sites of wind generation in **117**
Cornwall Council 41, 48–49, 52, 65, 82
COVID-19 pandemic 157–160, 167
cultural heritage value 42

Davidson D. 21
Death, C. 19
De Jong, E. 159
Department for Environment, Food & Rural Affairs (DEFRA) 41
Desertification Convention 17
Dey, I. 53
discourse: lessons for 173; sustainable development 173
displacement of belonging 143
dissolution 154–156, **155**
divergence 149–153
doughnut economics framework 34
Douglas, Mary 13

Earth Summit 17, 161
employment 106, 131; local-specific 101; predominant **102**; resource-based 101
Environment Act 2021 27, 35, 37
environmental: action 123; attitudes and behaviours 94, 133; awareness 50, 145; change 17; connections 145; conservation 92; decisions 33; degradation 150; despoliation 12; determinism 125; discourses 34; education 134; goal 19; governance 124, 160; heritage value 42; integrity 43; knowledge 50; limits 87; management 27; measures 17; organisations 139; policy-making 22; problems 13; projects 134; protection and regulation 36, 38, *76*, 78, 130, 139, 146, 157; rhetoric 159; risks 12–13, 16, 139; standards 37, 38; stewardship 106, 138, 160; sustainability 158; trends 17, 34; well-being 36
Environmental Impact Assessments (EIAs) 37
environmentalism 16, 103, 151
Escobar, A. 159
European Union 21, 153
evolution 156–160
expanded narrative 134–137

fishing 140–141
flood plains 85, *84*, 87, 89
Foundational Phase 160
Framework Convention on Climate Change 17
Frickel S. 21

Gallent, N. 43
general power of competence (GPC) 36
Giovannini, A. 29
Glass, L. M. 15
global: context, local governance in 27–44; and the local 20–21; and local connection 87–89, *88*; reflexive modernity connecting 164–175; sustainable development 164–175
Global Action Plan for the Earth 27–28
globalisation 11, 21–23, 130, 150; and role of nation state 21–23, **22**
global warming 4, 5, 11, 27, 38, 72, 78, 81, *83*, 87–88, 109; causes of *113*; Cornwall and 82–87, *83–86*; global and local aspects of 113; global scientific discourse of 112; narratives 111; uncertainty about 109–126
governance: capacity 49; frameworks 9, 12, 42, 50, 95, 149; in global context 27–44; legitimacy 4; local 60–69; local energy 39; and local sustainability 38; local sustainable development 20–21; mechanisms 72; multilateral 15–20; multilevel 32–33; participatory 30–32, *31*; process 3; processes 41, 48; role of local councils in climate 38–39; strategies 2; strengthening local climate 41; structures 14–15, 69, 72, 123, 147; sustainability 39
Great British Energy (GBE) 39
greenhouse gas emissions 86, *86*, 113
Green Prosperity Plan 39
Griggs, S. 29
Guiry, N. 22
Gupta, J. 15

Habermas, J. 21
Hargreaves, T. 40
Hass, P. 16
high-visibility industrial accidents 11

HIV/AIDS pandemic 150
housing 6, 39, 42–44; affordable 84, *84*, 98, 105–106, 113, 130, 165; and bank lending 94; and local intervention 42–44; in rural and coastal areas 42; social 87, 105
Hulme, M. 109

identity: coastal 140–141; community 58, 138, 139, 143
Institutionalisation Phase 160
internalising sustainable development 92–107
Irwin, A. 16

Jasanoff, S. 14
Jönsson, K. 159

Kanie, N. 159
Kemp, R. 156
Kvale, S. 53

labour government 2, 97, 104, 166
local: identifying to achieve global 147–161; and reflexive modernity 164–175; sustainable development 164–175
Local Area Agreements 79
local climate governance 41
local councils 32, 35, 145, 160; contextualising 27; in Denmark 40; in Germany 40; governance capacity 49; legal and financial autonomy to 38; and nature-based solutions 41; neighbourhood development plans 36; role in climate governance 38–39
local governance 60–69, *61, 62, 64, 66–68*; Climate Change Act 2008 37; community engagement 41–42; community-led renewable energy projects 40; comparative perspectives on 38; Environment Act 2021 37; evolution 33; in global context 27–44; historical context 33; local councils and climate governance 38–39; local government context 35; local intervention and housing 42–44; Localism Act 2011 36; and local sustainable development 32–33; local sustainable development policy 35; multilevel governance 32–33; National Planning Policy Framework 38; neighbourhood development plans 36; parish councils and cornish context 33–35; participatory governance 30–32, *31*; place-based climate adaptation strategies 40–41; strengthening local climate governance 41; Sustainable Communities Act 2007 36; UK centralised energy system and local disempowerment 39–40
local government 115; ability of 127; context 15, 35; in England 3, 65; first tier of 33; level 4, 38; organisations 44; restructuring 49, **49**; spaces of 29; sustainable development in 34
Local Government Act 1894 33
local intervention and housing 42–44
Localism Act 21, 27, 35, 36, 165
local sustainability governance 35, 38
local sustainable development 32–33; governance 20–21; policy 35

Macnaghten, P. 93
Millennium Development Goals (MDGs) 19, 166, 171
multilateral governance 15–20
multilevel governance and local sustainable development 32–33
Mythen, G. 28

National Association of Local Councils (NALC) 49
National Planning Policy Framework (NPPF) 38–39, 166
National Sustainable Development Strategies 22
nation state: globalisation and role of 21–23, **22**; in reflexive modern era **22**
natural resources 76, *76*, 77
neighbourhood development plans (NDPs) 36
Newig, J. 15
non-governmental organisations (NGOs) 17, 30

ocean 137–140
Organisation for Economic Co-operation and Development 154, 167

parish context 48–69, **55**; age *56*; ageing population *67*; complex needs *68*; consider county council goals *62*; Cornwall **48**, 48–49; county council considerate of local needs *62*; county

council support *62*; discourse analysis 52–53; gender *55*; grounded approach to theory development 53–54; interviews 52; lack of involvement *67*; linkages to county council *61*; local governance 60–69, *61, 62, 64, 66–68*; local government restructuring 49, **49**; occupation *56*; Parish demographics 55–60, *55–59*; partnership expanded 63–69, *64*; qualifications *57*; research phases and initial results 54–55; role in *58*; scattered layout of communities *66*; settlement type *59*; survey design 50–60, **51**, **55**, *55–59*; work in partnership *64*
parish councils 5, 27, 29–35, 37–44, 49, 52, 54, 55, 57–60, 63, 65–68, 72–78, 82, 85, 87, 88–90, 95, 120, 129, 134, 137, 145–146, 160, 165–169, 174; and the cornish context 33–35; emergence of 29; type **137**
Parker, G. 30
participatory governance 30–32, *31*
place-based climate adaptation strategies 40–41
planning decisions 33
Plymouth City Council 27
policy: lessons for 173; sustainable development 173
Pollinator Action Plans 41
Poortinga, W. 51
progress 13–14
Public Administration and Constitutional Affairs Committee Report (PACA) 3
public sector investment 2

Quality Parish 73; scheme 79; status 79

rationality 13–14
Ravetz, J. 14
reflexive framing **152**, 171
reflexive indicators 164, **168**; global 169, **170**; local 169, **170**
reflexive modernity 1–2, 4, 6, 12–15, 19, 22–23, 28–29, 33, 51–54, 60, 69, 72, **73**, 78, 89–90, 107, 116, 124, 127, 145, 148–149, 153–154, 156–159, 164–175
risks: salience of 109–127; and sustainable development 72–90
risk society 2; core principles of 11; expanding 11–13; governance 15; participatory governance in 30, *31*; and sustainable development 8–23, 27; synergies **12**; theoretical frame of 34
Risk Society; Towards a New Modernity (Beck) 11
Rosenau, J. 15

salience of risk 109–127
scepticism 96–106
science 13–14
Scoones, I. 159
second homes tourism 143
service provision, barriers to 65–69
Silverman, D. 53
Skerratt, S. 30
social housing 87, 105
Society of Local Council Clerks (SLCC) 49
socio-environmental futures 20
Starmer, Keir 2, 104
static shift of international discourse 157–160
Statistical Package for Social Sciences (SPSS) 52
Steiner, A. 30
Stern, Nicholas: *Stern Review* 111
Stern Review (Stern) 111
structural context of local sustainable development policy 35
survey 50; design 50–60, **51**, **55**, *55–59*; and reflexive modern indicators **51**
sustainable communities 6, 73, 79, 129–146; coastal identity and community 140–141; coastal isolation 141–142; community stewardship 144–145; community unpacked 129–134; expanded narrative 134–137; fishing 140–141; ocean 137–140; reclaiming the coast from within 144–145; second homes tourism and the displacement of belonging 143
Sustainable Communities Act 2007 36, 129, 165
sustainable development 1, 8–10, 12, *74*, *77*, *78*; in the beginning 92–96; community engagement in 41–42; connecting global to local 164–175; future research 173–175; global and the local 20–21; globalisation and role of nation state 21–23, **22**; governance structures 14–15; internalising 92–107; lessons for discourse and policy 173; local reflections 165–173, **168**, **170**, **172**; mechanisms *79–81*, 79–82; multilateral governance 15–20; narrative **161**; progress, science, and rationality 13–14; -related issues 31; and risk

72–90; and risk society 8–23; scepticism 96–106; within UN 147–149, **149**
Sustainable Development Goals (SDGs) 3–4, 6, 9, **10**, 15, 17–20, **18**, 18–19, **19**, 23, 95, 99–100, **100–101**, 153, 157–160, 166–167, 171, 174–175
Swyngedouw, E. 19–20

tourism: seasonal 42, 48, 93, 143; second homes 143; sustainable 42; tourism-driven economies 142
Transformational Phase 160
Turner, R. A. 34

uncertainty about global warming 109–126
UN Conference on Environment and Development (UNCED) 17
United Kingdom (UK): centralised energy system and local disempowerment 39–40; energy governance model 39; energy sector 39; Governments Public Attitudes Tracker 50; Sustainable Development Programme 21
United National buildings 150
United National General Debate Statement 149, 174
United Nations (UN): Sustainable Development Goals (UNSDGs) *see* Sustainable Development Goals (SDGs); sustainable development within 147–149, **149**; World Population Prospects 103
United Nations Conference on the Human Environment 16
United Nations Environment Programme (UNEP) 16, 153
United Nations General Assembly (UNGA) 18, 147, 166
Urry, J. 93

Vijge, M. J. 159
volunteer-led biodiversity conservation projects 31
Voss, J.-P. 156
Vries, B. J. de 158–159

Watkins, H. 30
wave energy projects 119, **119**
Whitmarsh, L. 81
Willett, J. 29
Wills, J. 2–3, 33, 97
World Bank 3
World Commission on Environment and Development (WCED) 8, 16; *see also* Brundtland Commission
World Conference on Sustainable Development 150
World Risk Society (Beck) 11
world risk society (WRS) 8, 14, 21, 23
World Summit on Sustainable Development (WSSD) 18, 166
World Trade Organization 21
Wynne, B. 112